Tim James is a secondary-school science teacher, YouTuber, blogger and Instagrammer. Raised by missionaries in Nigeria, he fell in love with science at the age of fifteen and refuses to get over his infatuation. After graduating with a Master's degree in chemistry, specialising in computational quantum mechanics, he decided to get straight into the classroom. His previous books are *Elemental: How the Periodic Table Now Explains (Nearly) Everything* (2018) and *Fundamental: How Quantum and Particle Physics Explain Absolutely Everything (Except Gravity)* (2019).

T0187001

Also by Tim James

Elemental
Fundamental

Astronomical

From Quarks to Quasars:
The Science of Space at its Strangest

TIM JAMES

ROBINSON

ROBINSON

First published in Great Britain
in 2020 by Robinson

5 7 9 10 8 6 4

Copyright © Tim James, 2020

ISBN: 978-1-47214-432-4

Typeset in Scala by Hewer Text UK Ltd
Printed and bound in Great Britain
by Clays Ltd, Elcograf S.p.A.

Papers used by Robinson are from
well-managed forests and other
responsible sources.

MIX
Paper from
responsible sources
FSC
www.fsc.org   FSC® C104740

Robinson
An imprint of
Little, Brown Book Group
Carmelite House
50 Victoria Embankment
London EC4Y 0DZ

An Hachette UK Company
www.hachette.co.uk

www.littlebrown.co.uk

For Bree

Contents

APPENDICES

'Equipped with his five senses, man explores the universe around him and calls the adventure Science.'

Edwin Hubble

In This Day and Age?

In 2016 the American rapper B.o.B. (real name Bobby Ray Simmons Jr) announced to the world via Twitter that he was a Flat Earther.[1] He also recorded a song about his beliefs, in which he criticised 'a cult called Science' for misleading everyone about how things in space truly work.[2]

While some ancient cultures believed Earth to be flat, the modern version of the Flat Earth movement was kick-started in 1838 when the English writer Samuel Rowbotham conducted an experiment to measure water levels along the Old Bedford River in Cambridgeshire. Rowbotham found that the water did not curve as much as he thought it should and thus declared Earth to be a disc rather than a globe. Naturally.

Rowbotham was, in all fairness, a skilled public speaker who could humorously outwit people who challenged his claims but soon his experiment was repeated by the scientist Alfred Russell Wallace, who took things like refraction into account and calculated that the Earth was round after all.[3] Even more telling perhaps was the version of his experiment carried out by Ulysses Morrow, who concluded the Earth was bowl-shaped, giving you an idea of just how reliable this experiment can be.[4]

A century and a half later, when B.o.B. started publicising these Victorian Flat Earth arguments, he had an advantage over Rowbotham: widespread fame and uncensored media coverage, without his theory being peer reviewed first. Within months of his declaration, other celebrities rallied to his cause and Flat Earth culture grew from an obscure fringe movement to a significant minority. So convincing are some Flat Earth arguments, in fact, that according to a 2018 YouGov poll 6.5 million Americans now believe Earth is not a globe.[5]

That figure might seem alarming but it shouldn't come as a surprise. Flat Earth arguments today have to come with an element of conspiracy (in order to account for satellite images) and who doesn't love a good conspiracy? Not only are conspiracy theories exciting and easy to understand, they make us feel smart for having seen a truth we aren't supposed to.

Mind you, I have never understood *why* the Illuminati would invent such a peculiar cover-up (not to mention how they persuade all the independent space agencies, airlines, pilots, GPS companies, navy personnel, mobile-phone industries, teachers, amateur astronomers and children with telescopes to go along with it), but that's beside the point. Conspiracy theories are fun and usually have exciting YouTube videos with spooky background music to support their claims.

I admit it can be frustrating as a science educator to deal with these Flat Earth arguments because we covered them during the Renaissance, but I also believe that when people have questions they should be allowed to ask them without ridicule.

In fact, the Flat Earthers I've engaged with have usually been erudite, sensible people and not the cousin-marrying yokels they are portrayed to be. They promote values such as scepticism and experimental evidence, which are, after all, true scientific values.

Obviously there are countless proofs that the Earth is round (see Appendix I for a few) but what interests me most about the Flat Earth movement is that all their arguments rely on the same approach. They point out an observation that does not seem to fit the globe theory, e.g. a star that doesn't move the way it should or a building that shouldn't be visible from a certain point, etc. and then ask: how does a round Earth explain *that*?

Some of the questions Flat Earthers ask are honestly quite reasonable, but the problem scientists face is that the answers are often so counterintuitive they can be hard to believe. Human brains are wired to handle straightforward things so when we come face to face with the Universe as it actually is, it can look . . . well . . . wrong.

Studying how things behave in space (astrophysics) or how the Universe evolves (cosmology) brings us face to face with scenarios so unusual they become downright disturbing. By definition of being 'everything ever', the Universe is the strangest collection of things

4

imaginable and, even when we know the facts, comprehending them is beyond our teeny mortal minds.

A Flat Earth view is simple and agreeable, but just because something appears obvious doesn't mean it's correct. In fact, in science the opposite is usually true. You only have to flick through a book of optical illusions to be reminded of how easily our simple everyday senses can be misled.

We live in a comfortable atmosphere governed by easily digestible laws of physics, but travel upward in a straight line and before you even reach an altitude of 10 kilometres (6.2 miles) the conditions become so different, so alien and so strange that your body literally starts to die. We aren't built to cope with the rest of the Universe so it's no surprise that when we look outwards, we find a cosmos peppered with weirdness and wonder.

Flat Earthers enjoy a manageable, safe view of reality but in order to learn about space properly we have to let go of instincts, intuitions and simple explanations. Those things are not required on a journey such as this. This is science at its absolute strangest.

PART I

OUR FREAKY UNIVERSE

CHAPTER ONE

Very Big, Very Old, Very Weird

The Astronomical Frontier

Any writer who tries to describe the magnitude of space is going to run into difficulty. The numbers involved are so extreme that some would say it's pointless to even try. But what kind of space book doesn't at least give it a go, eh?

First though, we need to appreciate the numbers involved in space physics. We throw words like 'million' and 'billion' around casually when we mean 'lots' but those numbers are actually very different. One million seconds is eleven and a half days for instance, whereas 1 billion seconds is thirty-one years (1 trillion seconds, if you're curious, is thirty-two millennia). Keep that comparison in mind while we try to wrap our heads around what's to come.

We'll begin by grappling with the fact that Earth is 150 million kilometres (93 million miles) from the Sun. Let's say you decided to fly towards the Sun in the world's fastest manned airplane, the Lockheed SR-71 Blackbird, which travels at roughly 1 kilometre (0.6 mile) per second. At that speed you could make the flight from London to San Francisco in two and a half hours.

Now imagine setting off for the Sun in that Lockheed on your eleventh birthday. Travelling at a constant speed without slowing for a second, you would be finishing secondary school by the time you reached your destination and that's only the distance in a straight line.

The Earth is currently ripping its way around the Sun thirty times faster than your Lockheed so, to get a sense of this, think back to what you were doing at this time yesterday. Whatever you were up to, you were doing it 2.5 million kilometres (1.6 million miles) away from where you are now. You're travelling fifty times faster than a bullet, fast enough to climb Mount Everest three times a second, and even at such a tremendous speed it still takes a whole year to make one orbit of the Sun.

The Sun is also no mere fireball. You could fit a million Earths into it, which is like filling one of those exercise balls pregnant women use with grains of rice – each grain representing an entire Earth. At that mass, the Sun is able to hold onto all the planets, right the way out to Neptune 4.5 billion kilometres (2.8 billion miles) away, a distance it would take 142 years to reach in your Lockheed plane.

Then, even further out, we come to the biggest and most distant structure of our solar system – the Oort cloud – a bubble of ice and rock encapsulating the Sun with a radius of 15 trillion kilometres (9.3 trillion miles). It would take a beam of light one and a half years to reach the Oort cloud from Earth, and 475,000 years for our Lockheed, longer than the human race has even existed.

The next nearest suns to ours are the Alpha Centauri cluster: a three-sun system comprising Proxima Centauri, Alpha Centauri A and Alpha Centauri B, which sit another four Oort cloud distances away. It takes light four years to reach them, which would be 2 million years for the Lockheed.

Further out, we get to other solar systems such as the awesomely named Wolf 359 and the slightly less awesomely named Lalande 21185, both of which take eight years for light to reach, 4 million for the Lockheed. And the distances get bigger still.

Suns clump together in disc-shaped clouds and when we look up at night, out in the countryside where there is no light pollution, we can see our own disc edge-on, looking like a ribbon of light stretching from one horizon to the other – a vapour trail made of stardust.

The Greeks believed this glowing band was milk squirted from the goddess Hera's breast to feed the baby Heracles abandoned on Earth. This is where we get the word galaxy, from the Greek *galaxias* meaning milky, and is also where we get the name of the galaxy itself: the Milky Way.

The size of the Milky Way has been a mystery for most of scientific history but in March 2019 a joint venture between NASA's Hubble Telescope and the European Space Agency's Gaia satellite was able to take readings of light density and cautiously weighed the galaxy in at 1.5 trillion times the mass of our solar system, containing roughly 200 billion suns. That's equivalent to the number of water droplets in a cloud. Which is pretty cool. We literally live in a star cloud.

That cloud is 100 quadrillion kilometres (62 quadrillion miles) from side to side, which takes light 106,000 years to cross. Our own solar system will reach the other side of its orbit around the galactic centre after 112 million years and we haven't even got to the silly numbers yet.

The nearest star cloud to us, the Andromeda galaxy, is 23 quintillion kilometres (14 quintillion miles) away and there is no way of putting that number in perspective. The closest we can maybe do is consider our own galaxy (which would take the Lockheed 3 billion years to cross) and say that Andromeda is 23,000 times further away, containing 1 trillion suns. And these galaxies are still only 2 of at least 100 billion others about which we know.

If you were to hold up a grain of sand to the night sky the area it covers up will contain at least 10,000 galaxies, each packed with billions of suns and Thor only knows how many other planets. Space is bigger than big, bigger than enormous, bigger than humungous, huge, vast, immense or colossal. The only word we can use to describe the size of space is 'astronomical'.

It's Also Very Old

Our best measurements about the age of the Universe clock it to about 13.8 billion years old, give or take a few thousand millennia. That's long enough to watch all five *Pirates of the Caribbean* movies in one go. But if that doesn't feel like something you could do, there is another good way of visualising the timescale.

Proposed in 1977 by the astronomer Carl Sagan, what we can do is compress the life of the Universe down to a single calendar year, with the beginning marked as midnight on 1 January and the present day becoming midnight on 31 December a year later.[1]

On this scale, the history of the human species (which in real time has lasted 200,000 years) takes up about four minutes. By contrast, the dinosaurs roamed for 170 million years, which translates to four days on the cosmic calendar – starting at lunchtime on Christmas Day and going extinct on 29 December as the leftovers were going off.

If we wind back to the very beginning of everything we would find that for the first few millennia everything looked very different because there were no atoms. At this point, the whole Universe was a glowing froth of free-floating electrons, neutrons, neutrinos, protons and photons. It was only after 380,000 years that things cooled enough for electrons and protons to combine and form the simplest atoms, hydrogen and helium. Or, compressed on our cosmic calendar, atomless energy settled into stable particles about fourteen minutes after midnight on 1 January.

These clouds of atoms, called nebulae, soon started to suck themselves inwards due to gravity and the nuclei of the atoms got crushed together, fusing them into heavier elements. The heat given off by these fusions was powerful enough to push back against gravitational collapse, and a balance arose between gravity pulling in and heat from the core pushing out. The resulting sphere of nuclear plasma was the very first sun – a furnace where light elements get roasted into heavy ones – and soon whole stellar nurseries were founded.

The oldest galaxy we know of is called GN-z11, observed in March 2016, and seems to have formed 400 million years after the Universe began;[2] mid-afternoon on 10 January of the cosmic calendar, roughly when everybody is wondering if they really do need that gym membership.

The oldest stars in our own galaxy indicate that it formed about 800 million years after the beginning – some time around 21 January – although there are only a few stars from this era left. Most of the first generation suns have long-since run out of atomic fuel and fallen apart, scattering heavier elements from their cores into the galaxy like fertiliser for the next generation of suns, which includes our own.

Our sun began knitting itself together 4.6 billion years ago (evening of 31 August on the calendar) but a lot of the elemental impurities made by the first generation of suns didn't make it into the furnace itself. Instead, they formed a giant dusty disc around the Sun, and these eddies and whirlpools of rock eventually became the planets.

It isn't just a wild guess that the planets congealed out of debris either: we can actually see it happening. There are a lot of suns out there to look at, so we can point our telescopes at the ones just coming together and, sure enough, we observe planets forming on the outskirts of every new star; planets such as Smertrios, which orbits the star HD 149026, or the less catchily named but easier to see PDS 70b orbiting the sun PDS 70.[3] These are planets in the process of birthing and we can watch every step in real time. All right, fair enough, technically they formed a long time ago and the light is only just reaching us today, but you know what I'm getting at.

The solar system we call home formed on midday of 1 September of the cosmic calendar, about 100 million years after the Sun started to glow. The four inner planets were stripped of their gaseous outer layers by solar winds while the outer planets were distant enough to keep their atmospheres, thus dividing the solar system into two regions of rocky inner planets and outer gas giants, with a circle of asteroids – remnants of a failed planet – in between.

Around 6 p.m. on this cosmic date, a tiny planet we have named Thea is thought to have wandered into Earth's territory and crashed into it. The resulting flotsam from this collision started orbiting Earth in its own right, eventually coalescing into another spherical structure a hundred million years after the Earth formed: the Moon.

For the next billion years or so, Earth was lifeless and scorched, until late evening on 30 September, when things became cool enough for the steam in the atmosphere to condense into oceans. Shortly after, the first primitive lifeforms starting squelching around in the mud. How this happened remains one of the most exciting mysteries in science, but we do know that the first plants began photosynthesising on 20 December with the aforementioned dinosaurs eating them five days later. Then, along came us, swaggering onto the scene at four minutes to midnight on 31 December.

If we now picture the final countdown everyone shouts as they approach the new year, the chant of 'ten' would represent the cradle of civilisation in Sumer followed by the pyramids being built at 'nine'. At five seconds to midnight, Jesus of Nazareth was born. Four seconds left and the city of Rome falls. Three seconds and Vikings start to conquer Western Europe. Two seconds and Genghis Kahn conquers Asia, forming the largest contiguous empire in history. Then, with one second left, Columbus lands on the shores of America and we are finally brought to the present day at midnight.

Looking to the future, we can extend our calendar and make solid predictions about what is going to happen to our solar system over the next cosmic year.

Three hundred million years from the present day – 8 January – Saturn's rings will have gone, having rained down onto the surface of the planet. But if you want a pretty ringed planet, worry not; 50 million years before Saturn loses its decorations one of Mars's moons, Phobos, will have been pulled apart by gravity and streaked into a circle around the planet, giving Mars a hula-hoop of its own.

Another 5 billion years and the hydrogen fuel at the centre of our sun will start to deplete. The Sun currently loses mass at a rate of 6 billion kg per second and once it has finished consuming its hydrogenous food, gravity will win and the outer layers will start to crumble.

Counterintuitively this will re-pressurise the core for a few extra millennia, giving it a final burst of furious fusion. The heat from this will be significantly hotter than anything that came before and it will inflate the outer layers of the Sun back outwards into space, turning it into a red giant, which will swallow Mercury, Venus and eventually Earth. By early May of next year on the calendar, our planet will be gone forever.

This outer envelope of solar material will dissolve over another billion years into the surrounding space, leaving behind a dense nucleus of glowing atoms, which was once its burning core: what we call a white dwarf star. White dwarfs are big, hot crystals of hydrogen, helium and a few heavier elements such as carbon and oxygen. Really, white dwarfs are big hot diamonds and I don't know why they aren't described like that more often.

The nearest white dwarf to us is Sirius B in the constellation Canis Major. It's much smaller than the Earth, but you can spot it with a decent telescope during the winter months (at least until the early 2020s when it will disappear behind Sirius A, the big star next to it). That faint pinprick of light is a forecast of what our sun is destined to become by the start of next cosmic June.

That crystal will slowly radiate heat away until it forms what we call a black dwarf, the cold corpse of a once blazing star. The time it takes for this to happen is longer than the age of the Universe, however, so no black dwarfs have actually formed yet, but eventually our sun will freeze and all that remains will be a lonesome black pearl. Which is, coincidentally, the name of the main ship in the *Pirates of the Caribbean* movies you will just have finished watching!

OH, AND IT'S WEIRD

There are two ways we can analyse other planets and stars. If they're inside our own solar system we can land probes on them and explore directly. At the time of writing we have landed fifteen times on Venus, sixteen times on Mars, sent two probes into Jupiter's atmosphere, one into Saturn's, landed on the surface of Saturn's moon Titan, put forty-seven probes onto our moon, three onto various asteroids, two onto comets and crashed one into the surface of Mercury. And when it comes to freakishness, our solar system does not disappoint.

Mercury has a thin atmosphere of helium on its surface and an asteroid crater big enough to swallow Texas. It also has tails of hydrogen gas coming off its surface like plant shoots reaching for the Sun – nobody knows why.

On Venus, the hottest planet, reaching temperatures of 460 °C, the rain is made of concentrated sulphuric acid, which boils before it hits the ground and flies back upward. The atmospheric pressure is over ninety times greater than our own so walking on the surface of Venus would be like walking on the bottom of the ocean, and it snows pellets of lead sulphide rather than ice.

Mars, a planet populated entirely by human-made robots, is coated in a fine layer of red rust with liquid water stored in giant lakes beneath the surface. Its daytime skies are pink, its sunsets are blue, it is home to the tallest mountain in the solar system (Olympus Mons – two and a half times taller than Everest), and its climate features avalanches and dust storms so big they cover the whole planet.

Jupiter, twice the mass of the other planets combined, is also the fast-est spinning, with its day lasting only ten hours. The core temperature is estimated to be around 24,000 °C, six times hotter than the surface of the Sun, and it generates the largest object in the solar system: a magnetic field twenty times stronger than Earth's.

Jupiter also features the famous red spot, a storm twice the width of Earth whose colour origin is a mystery, and plays host to about eighty moons such as Io, the only other place in the solar system with active volcanoes, and Europa, which has a giant ocean of liquid water below an icy shell. Oh, and Jupiter is the only other planet to have LEGO™, because the Juno probe that flew there in 2016 had three figurines of Galileo, Jupiter and Juno stashed inside.[4]

Saturn, with its rings of ice and rock less than a kilometre thick, is so fluffy that if you were to somehow build a bathtub big enough, the entire planet would float on the surface like candy floss. Over sixty moons orbit it, including Titan, which has rivers of liquid methane (the fuel used to power school Bunsen burners) flowing over a landscape of solid water.

Uranus is a giant ball of ice, mostly composed of water, ammonia and methane, but it has a hydrogen/helium outer atmosphere, which gives it a blue tint. It is the only planet to orbit the Sun on its side (probably hit by another planet in the distant past) and it has close to thirty moons, named after characters from Shakespeare plays.

Neptune, the farthest planet, is also the windiest, with 2000 km/hour (1243 mph) hurricanes racing across its surface (nobody knows how this is possible). It is four times the size of Earth but far less dense, meaning Neptune's gravity is almost the same as our own, so we would actually feel more gravitationally comfortable on Neptune than on Mars. Neptune also has the most unusual rain in the solar system because its atmospheric conditions are just right to crystallise carbon snowflakes in the upper atmosphere to form tiny hailstones made of diamond.

And that's just our neighbourhood. If we want to observe planets far away, however, we have to do so using a technique called spectroscopy, invented by Robert Bunsen (the aforementioned burner guy). It works on a simple principle: different chemicals emit and absorb different frequencies of light so if we analyse the light from distant objects we can figure out what they are made of.

When a planet moves in front of a star we are studying, the star light passing through the planetary atmosphere gets partially absorbed depending on what kinds of chemicals are present. By noting which frequencies get absorbed by the obscuring planet, we can thus figure out what the weather is like on a far-off world. And when it comes to freakishness, the rest of the galaxy does not disappoint either.

Consider the planet 55 Cancri e, which is believed to be made of one-third diamond right to its core.[5] Or the planet J1407b, which has rings like Saturn only stretching 640 times farther out, meaning the planet looks like a pea at the centre of a dinner plate.[6]

There's a planet called Wasp-12b, as black as asphalt and egg-shaped due to a slow elongation process pulling it towards its sun (it has 10 million years left).[7] Then there's CoRoT-7b, which is so close to its sun it has whole oceans of lava on its surface, or take Gliese 1214 b, which is believed to be made of water pressurised solid, but so hot it catches fire, making it a planet of flaming ice.[8]

There's a nebula called Sagittarius B2 composed of ethyl methanoate, the chemical that gives raspberries their flavour.[9] There's a solar system called Castor composed of six suns weaving in and out of each other like a juggling trick, and there's a star called V Hydrae that ejects cannonballs of plasma twice the size of Mars into space.[10]

There's a planet called TrES-2b, which is the darkest in the galaxy, absorbing almost all the light which hits it.[11] There's a planet named OGLE-TR-56b where it rains molten iron.[12] There's a planet named HAT-P-7b, twice the size of Jupiter, where the rain is made of aluminium oxide, the main chemical component of rubies.[13] And then there's HD 189733b, a planet where it rains molten glass. Sideways.[14]

Space is very big.

Space is very old.

Space is very weird.

CHAPTER TWO

One Dead Moose

Water and Lava

Every human culture in history has catalogued the stars. Perhaps because astronomy reassured our ancestors that no matter what chaos they faced on the ground, the sky was unchanging and stable. Or maybe it was the simple fact that stars are gorgeous and it's fun to gaze up at the sparkly darkness. Whatever the reason, everyone seems to have discovered that the constellations follow a repeating pattern over the course of a year, which is handy if your civilisation is learning to do agriculture and needs to predict when the seasons are about to change.

This discovery that we can predict the future from studying the stars was championed by the Babylonians and it lingers on today in the form of astrology and horoscopes. Sadly, astrology's predictions are usually quite vague (see Appendix II) so if you want something sophisticated you'd have to talk to the ancient Greeks.

Before the Romans invaded and killed everyone, Greek astronomy was remarkably ahead of its time. Anaximander (590 BCE), for instance, realised that since Earth was spherical the sky had to continue below its horizon, meaning the atmosphere is a bubble around the planet, rather than a dome. Or take Anaxagoras (500 BCE), who thought stars were made of fire, or Aristarchus (280 BCE), who speculated that the Earth orbited the Sun rather than vice versa.

The person who made the first precision prediction, however, was Thales of Miletus, who predicted a solar eclipse on 28 May 585 BCE. Based on careful study of historical records, Thales realised that eclipses were not the result of gods being angry – they followed a rhythm, just like the stars. He didn't know what caused them (that was figured out by the Chinese astronomer Shi Shen a few centuries later) but he knew there had to be a mechanism behind what he was seeing.

Thales is also on record as having stated that everything in existence, including himself, was made of water. That sounds a little kooky but his reasoning was that since water takes so many forms (snow, ice, hail, steam, fog, etc.), everything was probably derived from water in one way or another.

The spirit of science is encapsulated perfectly in these two facts about Thales. When he tried to figure out how the world worked, using imaginative ideas, he ended up stating something ridiculous. When he based his thinking on testable predictions he ended up making an important discovery. Science is not just about having clever ideas, it's about checking to see if they're right.

For my money, of course, the most committed scientist of the Classical era was a man named Empedocles who lived during the fifth century BCE. Empedocles had a wonderful knack for getting every hypothesis he came up with totally wrong but that didn't hold him back.

Empedocles suggested that there are four chemical elements. There are 118. He suggested that stars are fixed to crystal domes revolving around us. They aren't. He suggested that our eyeballs shoot magical lasers, helping us to see. They don't. And most importantly, he claimed to have immunity to lava and that he would be able to jump into the mouth of an active volcano and survive. He didn't.

ARE WE MOVING?

Thales's theory that everything in the sky could be predicted by underlying laws worked for phases of the moon, cycles of constellations and eclipses, but there were five objects in the sky that didn't fit. These five points of light zigzagged about, sometimes looping back on themselves in what we call retrograde motion.

The great philosopher Plato challenged his students to come up with a law that might reveal a pattern to these movements, but nobody could do it because their paths didn't seem to follow anything reliable. Most Greeks decided therefore that each one was controlled by one of the gods and gave them the names Hermes, Aphrodite, Ares, Zeus and Chronos, collectively referred to as *planetes*, which means 'wanderers'.

A few centuries later, during the period of their aforementioned imperial expansion, the Romans decided they liked the idea of planets but didn't want to keep the Greek names. So they renamed them Mercury, Venus, Mars, Jupiter and Saturn. Then everyone died, and astronomy was squashed for about a thousand years during the dark ages.

Fortunately, the Arabic world had a slightly more enlightened view of things and continued to make progress in investigating the sky, leading to a very curious discovery made by the scholar Abu Sa'id al-Sijzi some time during the tenth century. We know very little about al-Sijzi as a person but we know he is credited with inventing a type of device called an astrolabe, which simulates star movements.[1]

Other astronomers had been building mechanical star charts for at least a thousand years (the Chinese astronomer Geng Shou-chang might have built one as early as 70 BCE[2]) but al-Sijzi's was unique because, in his design, he kept the stars fixed and rotated the Earth instead, which, to everyone's surprise, yielded identical predictions.

Assuming the Earth is stationary was apparently just that – an assumption. There was no actual evidence that said it had to be that way round and switching to the opposite perspective made no difference. Thanks to al-Sijzi we could no longer prove the Earth was still.

But surely if the Earth *were* the thing that was spinning, we would feel wind resistance against us constantly. Also, jumping into the air would bring us down in a different place as the planet turned beneath our feet. Al-Sijzi's astrolabe results had to be a coincidence, right?

Well, unfortunately not. As Bishop Nicole Oresme pointed out in 1377, these common-sense points do not actually work.[3] The Earth's atmosphere would be moving with the Earth at the same rate so the air would appear motionless relative to people on the surface. Also, jumping up and down would not displace you, for the same reason it doesn't displace you if you jump up and down inside a train.

When you jump from rest you are only at rest relative to the floor, which is already moving with you. From the perspective of someone outside, you would be seen to jump in a parabola, relative to the ground. Only *you* see things as vertical. Likewise, anyone standing in space watching you jump on the surface of the Earth would see you jump in a parabola, relative to the stars. Al-Sijzi's discovery was therefore valid. There was no way of deciding whether the Earth was moving or not.

Idiots of the Renaissance

Quite a few solutions to the planet problem were offered in the ensuing centuries, usually involving complicated things called epicycles, where each planet was assumed to be going round on a separate circular path of its own, while also orbiting the Earth.

But then in the early sixteenth century something radical was suggested by the Polish academic Nicolaus Copernicus. Copernicus was a brilliant man who had studied both medicine and law at university, as well as being a canon of Frombork cathedral.

Just before he died in 1543, Copernicus published a book in which he stated that we had everything backwards.[4] The Sun was at the centre of the solar system and the planets, including the Earth, were orbiting it. This would explain the peculiar retrograde motion of planets because as we went round, we would sometimes overtake the planets further out. At certain times they would appear in front of us, but when we caught them up and overtook them, they would appear to move backwards.

Copernicus also believed Earth was the third planet from the Sun, with Mercury and Venus closer, and showed that if we made the assumption of a heliocentric (sun-centred) universe, the data would make more sense and we wouldn't need to include countless epicycles. After writing this incendiary book, Copernicus promptly died, leaving his suggestion hanging in the air like a belch at a wedding.

It isn't clear exactly why he waited until his imminent death to publish his idea, although it might be because heliocentrism would be in conflict with a literal interpretation of the Bible, which states that the Sun is in motion, not the Earth (Joshua 10:13, Ecclesiastes 1:5). Asking questions was one thing, but contradicting the Bible was a pretty unwise thing to do in sixteenth-century Europe.

It has to be said, though, that contrary to popular myth, Copernicus's idea was not challenged by the Church for a long time, although that was mostly because nobody took it seriously. The idea of the Earth going round the Sun was veritable poppycock and even the great Martin Luther mocked the idea, saying (paraphrased): 'So it goes these days. Whoever wants to appear clever must disagree with what everyone else says. He must do something of his own. This is what that fool does, wishing to turn the whole of astronomy upside down.'[5]

Copernicus's model did not make any new predictions, mind you; it was just mathematically a lot easier to use than anything else at the time. As a result, astronomers began adopting it purely as a calculation tool and, in fairness to them, there was no reason to treat it as anything more. It did not *quite* match the measurements, so while you could use it for quick and dirty calculations, you had to use the standard 'Earth is not moving' approach to get the best results.

The person who did the most work on boosting the Copernican method (without actually agreeing with it) was the Danish nobleman Tycho Brahe, arguably the greatest astronomer of his age as well as one of its most notable weirdos.

Brahe was born into a wealthy family, the eldest of twelve children, three years after the death of Copernicus. His father was one of King Frederick II's chief advisors and Tycho was expected to join the politics business, but during his teen years, he became hooked by science and decided to pursue that instead. Good move.

Tycho was such a passionate scientist that he once fought a duel with an academic named Manderup Parsberg over who was better at equations. In the process, he lost his nose to Parsberg's sword, forcing him to wear a bronze prosthetic for the rest of his life.

Tycho's luck changed, however, when he saved King Frederick from drowning a few years later, and he was rewarded with a castle on the island of Hven. Not only that, King Frederick furnished the castle with an astronomical observatory, a chemical lab and a personal printing press. Essentially, Brahe had his own science theme park, which he named Uraniborg after Urania, the Greek muse of astronomy.

Slightly more bizarrely, Brahe hired a telepathic dwarf called Jepp to live with him and follow him around singing songs all day. He also kept a pet moose on site with him for reasons that have never been clear.

Unfortunately, the moose met with a tragic and suspicious death when it fell down a staircase. Brahe's defence was that it wasn't his fault because the moose was drunk at the time and shouldn't have been attempting to go down the stairs anyway.[6] Although that's probably the excuse I'd have given as well if I was being accused of moose murder.

During his time at Uraniborg, Tycho Brahe made major contributions to astronomy, such as being the first person to realise stars were further away than the Sun, but his greatest work arose from a tense collaboration with the German astronomer Johannes Kepler.

Kepler was a big supporter of the Copernican approach and felt that, although the numbers didn't match perfectly, the idea was more elegant than putting Earth in the middle. He made a living as the chief astrologer to Emperor Rudolph II and wrote horoscopes for him based on astronomical data provided by Brahe. Kepler himself did not believe in horoscopes, of course, but he sold out because, hey, you gotta pay the bills somehow. By the way, eat Pringles™.

Kepler and Brahe famously disliked each other, possibly because Kepler was a notoriously shy man and Brahe was the kind of person who would get a moose drunk and throw it down a staircase while his psychic dwarf sang to him. Brahe would also hog the data sometimes and not let Kepler into his lab to do calculations, so the two of them became bitter collaborators.

In fact, Brahe's death was quite suspicious in itself, and according to some legends he was murdered by Kepler via mercury poisoning. That is speculation, but what *is* known is that after he died, Kepler nabbed all his data and took it back to Germany in order to prove Copernican astronomy right.

Once Brahe wasn't obscuring his progress, Kepler finally managed to show that the Copernican model *did* match the data, provided you made one adjustment: the planets were not orbiting in perfect circles but in ellipses.

Once this assumption was made the heliocentric theory turned out to be every bit as good as the geocentric one. Kepler was pretty acerbic about it too. His book on the topic opens with the following paragraph . . .

'Advice for idiots: Whoever is too stupid to understand astronomical science or too weak to believe Copernicus without affecting his faith, I would advise him [to] mind his own business and take himself home to scratch in his own dirt patch.'[7]

THUNDERBOLT AND LIGHTNING, VERY VERY FRIGHTENING, ME[8]

Galileo Galilei was born in 1564 and was well into his thirties by the time Kepler published his heliocentric theory of the solar system. In secret, Galileo wrote to Kepler and endorsed the theory, saying that he too favoured a Copernican view but was too frightened to come forth about it.

It wasn't until 1609, eight years after Tycho Brahe's death, that he decided to have a go at building his own telescope and point it towards the sky. Telescopes were a well-known instrument, likely invented in the Arabic world, but Galileo was able to improve on the design and built a version that magnified distant objects by a factor of twenty.

Over the next few years, he made a number of important observations including but not limited to . . .

1 Mountains on the Moon only glow when sunlight hits them, showing that the Moon is reflecting light rather than producing it.
2 The Milky Way is made of stars and not breast milk.
3 Saturn has rings, which he described as 'ears'.
4 Jupiter has four moons, undermining the belief that the Earth is special.

Then, in September 1610, he made (and I say this without a trace of hyperbole) one of the most important discoveries in scientific history.

The phases of the moon are the result of sunlight hitting it at various angles. This can be explained using the geocentric idea without much fuss. But Galileo discovered something that should be impossible: Venus has phases as well.

The geocentric system, with all its rotating domes and epicycles, can't account for that. It could point to where Venus was but explaining its phases could only be done if Earth was orbiting the Sun and we were seeing different bits of Venus at different times. Suddenly the Copernican theory was the only one that could explain the data and, at this point, perhaps ironically, the story got pretty dark for Galileo.

Up until then, the Church had tolerated Copernicanism. With Galileo's discovery that it was correct, however, things changed. Cardinal Robert Bellarmine wrote of Galileo's discovery: 'To affirm that the Sun really remains at rest ... and that the Earth is situated in the third heaven ... is a very dangerous thing. Not only may it irritate all the philosophers and scholastic theologians, it may also injure the faith and render holy scripture false!'[9]

Galileo did not help matters by writing to his friend, Christina of Lorraine, explaining that the Bible should not be taken literally and that scripture 'tells us how to go to heaven, not how heaven goes'.[10] He even went all the way to Rome and argued the case for Copernicanism to the Church. Unfortunately, the tribunal hearing his statements dismissed them and declared his argument 'foolish, absurd, heretical and contradicting Holy Scripture'. The fact that he had evidence on his side did not matter. The Bible was infallible and had to be interpreted literally. That was the end of it.

For some reason, Galileo then made the slightly unwise decision of relocating to Florence. He had previously been working in Padua, a city not subject to Vatican rule, but in Florence he came directly under the power of the Roman Catholic Church and was summoned before the Inquisition.

The Inquisition secretly instructed him not to teach Copernicanism, and publicly banned any works that contained reference to it. Galileo followed these instructions obediently, until his friend Maffeo Barberini was elected to become Pope Urban VIII in 1623. Galileo hoped his friend would give him support, but unfortunately the new Pope refused to overturn the Inquisition's decision.

Galileo's response, naturally, was to write a novel in which the main characters explain Copernicanism to an idiot named Simplicio (which translates as 'Simpleton') modelled directly on Pope Urban VIII, even quoting him in his dialogue. Galileo thus publicly contradicted the Bible. And insulted the Pope. In Italy. During the Inquisition. In many ways, Galileo was one of the finest minds in history. In many ways, one of the dumbest.

The Inquisition banned Galileo's novel and, in a move that shocked absolutely no one, put him on trial. Initially, Galileo was defiant and refused to revoke his position . . . until he was presented with torture devices, at which point he broke down and made a plea bargain, confessing that he had let arrogance go to his head. He further agreed that he had been wrong to assert Copernicanism and remained under house arrest until he died in 1642.[11]

Eventually, 337 years later, Pope John Paul II requested the trial of Galileo be reopened and, after carefully reviewing it, he issued a formal apology, saying that the Church had been mistaken in its handling of the case.[12]

If there is one important lesson to be learned here, it's that even the subtlest measurement can bring an established theory crashing to the dust. This can sometimes bring scientists into conflict with non-scientists because it looks like we're trying to be smart alecs, challenging treasured beliefs. The counterargument is that if a belief can be broken so easily, it cannot have been that stable to begin with.

Fortunately, when astronomers present something that overturns a long-held belief nowadays, it no longer results in court cases and torture threats. Just a lot of angry people on the internet . . .

IT'S ALL FUN AND GAMES UNTIL SOMEONE LOSES A PLANET

In 2006 the International Astronomical Union (IAU) declared that Pluto was no longer a planet and henceforth would be referred to as a 'dwarf planet'. The man responsible for the decision, astronomer Mike Brown, uses the Twitter handle @plutokiller and has a banner picture of Alderaan being destroyed by the Death Star in *Star Wars: Episode IV – A New Hope*. I believe you call that 'trolling'.

Obviously nobody likes to be dictated to, so it's no surprise that people got annoyed about this decision. Don't get me wrong, public opinion should not dictate truth because facts are not a democracy (they are a dictatorship in which nature tells us how she is and we comply), but definitions of words *are* democratic and if people want to use 'planet' to describe Pluto, shouldn't they be allowed to?

Well, not in this case. What this viewpoint gets wrong is that it misunderstands the IAU's motivation in reclassifying Pluto. In actuality, they *were* taking public opinion into account and the reclassification of Pluto was done out of respect for the lay public, not in spite of them.

The first definition of planet was obviously 'one of them five weird lights in the sky', but by the end of the Renaissance we had figured out there were six worlds going round the Sun, one of which we were riding. The definition of planet thus became more sophisticated and meant: object orbiting the Sun.

Then, in 1781, William Herschel discovered that one of the dimmer objects in the night sky, previously thought to be a star, was actually showing retrograde motion, making Herschel the first person in recorded history to discover a planet. There were now seven: Mercury, Venus, Earth, Mars, Jupiter, Saturn and George.

Herschel named the seventh planet George to honour the King of England, George III, but this did not catch on in France and it was renamed after the Roman god of the sky, Uranus. Now, look . . . as a physics teacher I have probably heard every permutation of the Uranus joke possible. But I've got to be honest, Uranus does make me laugh.

In 1801, Giuseppe Piazzi discovered an eighth planet lurking between Jupiter and Mars which he named Ceres. Then, a few months later, Heinrich Olbers discovered another planet along the same orbit, which he named Pallas. In 1804, Karl Harding discovered the planet Juno, in 1807 Heinrich Olbers discovered Vesta and in 1845 Karl Hencke discovered Astraea.

The next one was a bit different, though, because it was the first planet to be discovered by equation rather than telescope. A few decades prior, Alexis Bouvard was taking precise measurements of Uranus (hur hur hur) and found it did not move in a perfect ellipse. It seemed to be pulled off to one side, as if there were another planet gravitationally attracting it. Wind forward to 1846 and Johann Galle confirmed the existence of this thirteenth planet: Neptune.

The following year Karl Hencke discovered the planet Hebe, once again between Mars and Jupiter, then John Hind discovered Iris, followed by Andrew Graham's discovery of Metis in 1848 and Annibale de Gasparis's discovery of Hygeia in 1849.

Any bog-standard 1860s astronomy book would have proudly listed the seventeen known planets but as our telescopes got better we kept discovering more and more orbiting between Mars and Jupiter.[13] Hundreds in fact. Which presented a problem.

When people thought of the word 'planet' they imagined huge spherical worlds ploughing across lonely orbits – not scraggly chunks of space lint forming a loose belt around the Sun. We either had to keep the word planet meaning 'thing which goes round the Sun', or we had to modify it to match what people actually meant when they said it. We decided to go with the second option and introduced the word 'asteroid' to describe the stuff between Mars and Jupiter, because when people said 'planet' they meant big lonely round things. You can probably see where this is going.

In 1906 the millionaire astronomer Percival Lowell had begun to suspect there was a ninth planet past Neptune because it wobbled just like Uranus (hur hur hur). Sadly, Lowell died before his Planet X was discovered, but the Lowell observatory continued his work. Under the direction of chief astronomer Vesto Slipher, a young man named Clyde Tombaugh was set the task of searching the outer solar system and on 18 February 1930 captured the first image of a ninth planet, roughly the size of Earth.

Planet X fever hit the headlines and an international competition was organised to decide what to call this new world. Pluto was proposed by eleven-year-old Venetia Burney and won by popular vote.[14] However, by 1948 more precise measurements on Pluto were carried out and we realised we had overestimated its size somewhat. Pluto was only about a tenth the mass of Earth. Not as impressive, but oh well, it was still big enough to match what we think of as a planet.

Except it wasn't. By 1978 we had learned that Pluto was actually a six-hundredth the mass of Earth, smaller even than the Moon. Then, in 1992, the astronomers Jane Luu and David Jewitt discovered another object floating along Pluto's trajectory, nicknamed planet Smiley (its official designation is 1992 QB_1).[15] Then, in 2003, Mike Brown discovered another object nearby, which he called Sedna, followed by Haumea, Orcus, Makemake and, in 2005, Eris, 27 per cent heavier than Pluto.

It turns out there are over 2000 objects floating beyond Neptune, and Pluto is only one of them. Our solar system doesn't have a single asteroid belt, it has two. The second is named the Kuiper belt and that presented us with the same problem as before. Pluto was *not* what people thought of when they imagined a planet, it was just a fat asteroid. A fatsteroid, if you will.

If we kept calling Pluto a planet we would be misleading people about what it was. But if we redefined the word planet, a lot of asteroids in the inner solar system would suddenly be upgraded and we would have a similar public outcry. Instead of people complaining that Pluto was no longer a planet we would have moans of 'Ceres isn't a real planet!' etc.

Eventually the IAU decided that the definition of a planet was fixed in people's minds. A planet is a) something which goes around the Sun; b) is heavy enough to have gravitationally pulled itself into a sphere; and c) has cleared its orbit path so it's the only dog in town. An object that only fits the first two categories is just a dwarf planet, of which there were five known: Ceres, Pluto, Eris, Haumea and Makemake.

This made a lot of people angry, although I think that anger is misplaced. Children to whom I teach astronomy are fine with Pluto not being a planet, because they have grown up with the fact that 'Pluto was accidentally mislabelled' as part of what they learn.

It's not that Pluto was demoted, we just discovered what it was all along, like taking the mask off a Scooby-Doo villain. I get why people were not happy – nobody likes having a childhood fact removed – but unfortunately that's part of growing up. Science is tough sometimes.

For Real this Time?

In 2014 Chadwick Trujillo and Scott Sheppard were taking routine measurements on asteroids in the Kuiper belt when they spotted something tantalising. Some of the asteroids were not orbiting at the expected angle.

If you imagine looking at the solar system side on, most pictures and models portray everything flat, like a disc cutting through the Sun's equator (called the celestial plane). In reality this is not what happens. Objects orbiting the Sun usually do so at a tilt from the celestial plane.

Most planets have small tilt angles but when we get to the Kuiper belt objects are less affected by the Sun and orbit at bigger deflections. The degrees of tilt should be random, but Trujillo and Sheppard found a group of objects bunched together, orbiting in a neater pattern than expected, which suggests something is out there beyond the Kuiper belt.[16]

Then, in October of 2018, Trujillo and Sheppard discovered something else: a new dwarf planet, which they named the Goblin since it was discovered near Halloween.[17] The Goblin, like the earlier group, orbits as if something large is pulling on it. At the time of writing, nobody knows what is causing these anomalies, but one of the most likely explanations is that our solar system has a ninth planet after all.

Early calculations suggest it could be up to ten times the size of the Earth, which is very surprising. Our current understanding of planetary formation makes it unlikely that such a large amount of matter could have gathered so far out. If there really is a 'Planet 9' and it really is as big as we think, either our model of planetary formation is wrong or Planet 9 is a rogue planet ejected from its own solar system.

Or, if you want a more extreme hypothesis, it's possible Planet 9 isn't even a planet at all but something far more terrifying. According to the theoretical physicists Jakub Scholtz and James Unwin, the object pulling on the Kuiper belt could be a primordial black hole – a black hole the size of a watermelon with five times the mass of the Earth, formed billions of years ago, near the beginning of the Universe.[18] To be clear, Scholtz and Unwin are not claiming that's what it is, they're just saying we shouldn't rule anything out at this point.

Spotting whatever this object is, if it's there at all, is going to be difficult because planets and black holes do not generate light and are only visible when reflecting or deflecting light from the Sun. That's why planets don't twinkle, in fact; their light is closer to us and therefore more intense than the faraway stars, i.e. it's less likely to be bounced around in the air currents of Earth's atmosphere, which is how twinkling originates.

This means the further away a planet is the less light it reflects and thus the harder it is to see. Space is somewhat dark, after all. One thing is certain, though. This object, whatever and wherever it is, is likely to be bigger, darker and less hospitable than Uranus.

CHAPTER THREE

Party Cake and Pigeons

The Sirens Are Calling

Ask anyone to do an impression of a car racing past them on a motorway and without doubt they will make the following noise: Niiiiieeeeaaaaaoooooowwwww. Don't pretend you didn't hear that in your head.

Sound waves are compressions and expansions in the air which emanate from a vibrating object. Compressions far apart are said to have a long wavelength, which our brains interpret as a low-pitched note, while compressions bunched together have a short wavelength and a corresponding high-pitched note. Where this gets interesting is if the vibrating object is moving.

If you imagine standing in front of a car as it approaches you, the vibrations in the air become squashed together since the car is moving into its own sound wave. Your eardrums subsequently pick up lots of compressions per second and you hear a high pitch.

If you stand behind the car, however, the pulses appear stretched apart because the object is moving away from its sound wave and your ear will interpret this as a lower pitch. You sometimes notice this effect with ambulances and fire engines when their nee-naws seem to 'droop' as they go past you at high speed.

This phenomenon is called the Doppler shift and can be observed for any kind of wave. Since beams of light behave like waves transferring energy from one point to another, moving objects will not only sound different, they will *look* different when they move towards or away from you.

If I were to throw a glowing light bulb towards you, the light wave reaching your eyes would be slightly squished together, giving a 'higher pitched' beam of light – what the eye perceives as blue/violet. If you throw it back to me, the light wave reaching your eye is now being stretched instead, creating a 'lower pitched' beam of light – the colour red.

Unlike the obvious pitch changes of sound, though, the Doppler shift for light is extremely subtle and your eye is not sensitive enough to detect it. The distance between adjacent light pulses is in the order of a millionth of a metre, which means you won't notice the effect unless the object is moving very fast. At interstellar speeds.

RED-LIGHT DISTRICT

In 1912, Vesto Slipher (the guy who ran the project that discovered Pluto) was analysing light from distant galaxies to determine their composition. What he noticed was that other galaxies are giving off a light signature identical to that of the element hydrogen, but shifted towards the red end of the colour spectrum. Hydrogen is the most abundant element in our own galaxy so it made sense that we should see it elsewhere. But why was the light skewed towards red?

For a long time nobody understood what this observation could mean but in 1927 the Belgian priest and physicist Georges Lemaître offered a bold suggestion: what if red-shifted light was the result of other galaxies moving away from ours? What if the Universe was in a process of expansion and light waves from distant galaxies were becoming stretched by their movement?

Einstein read Lemaître's paper and was impressed with the mathematics, but felt that the idea of an expanding Universe was too silly and, at least according to physics legend, wrote to Lemaître saying 'your maths is good but the physics is atrocious'.

Later that same year, the American astronomer Edwin Hubble announced an even more puzzling discovery: the further away a galaxy is the more red-shifted it appears. This has all sorts of peculiar implications so let's take a close look at how this could arise.

Imagine seeing a particular galaxy, let's call it Arnold, and measuring its speed going away from us as 100 metres per second (m/s). Now imagine spotting a more distant galaxy beyond Arnold, let's call it Belinda, which is also moving away from us at 100 m/s. Both Arthur and Belinda are retreating from our vantage point at the same speed, like two wheels on a tandem bike, so their light should be red-shifted by the same amount.

But now suppose you lived in the Arthur galaxy and pointed your telescope towards Belinda. Belinda would appear stationary to you since you are moving at the same speed as each other. But if you pointed your telescope towards the Milky Way, you would see it moving away at a rate of 100 m/s. You would logically conclude that your galaxy and Belinda were the two central galaxies of the Universe and everything else was flying outwards from your location. But that is not what happens.

Hubble found that Belinda, the farther-out galaxy, is actually moving away from us faster than Arthur is, let's say at 200 m/s. An observer in the Arthur galaxy would therefore see Belinda moving away from itself at 100 m/s and the Milky Way moving in the opposite direction also at 100 m/s. In other words, if everything is moving away from everything else, the farther away something is, the faster it appears to be retreating.

There are a few exceptions to this red-shift phenomenon, e.g. the Andromeda galaxy is slightly blue-shifted (meaning it is on a collision course for the Milky Way and will arrive in 4 billion years), but the picture is, overall, unambiguous. The farther away a galaxy is the faster it appears from our perspective, which means every galaxy is moving away from every other galaxy.

This discovery is exactly what you would see if the Universe were expanding, and as soon as he learned of Hubble's discovery, Einstein conceded he had been mistaken and immediately began promoting Lemaître's expansion idea. And it never hurts to have the endorsement of someone like Einstein.

LIVING LEGEND

There are just as many myths about Albert Einstein as genuine facts. I reckon one of the reasons for this is that while everyone knows he was a genius, not many people can articulate what he actually did, so myths get invented to fill the gaps. And he did a lot.

Among his many contributions to science were the first proof that atoms existed, the earliest version of quantum theory (for which he won the 1922 Nobel Prize), the discovery that energy and mass can be inter-converted via $E = mc^2$ and the discovery that the faster you move through space, the slower you move through time – part of a theory called special relativity.

He also invented a refrigerator without moving parts and, more excit-ingly, a gentleman's shirt with an adjustable gut-strap to contain your paunch![1] Sadly, he did not get a Nobel Prize for this, proving there is no justice in this cruel, cruel world.

His grandest invention, however, the one for which he is most vener-ated, was the one he published in 1915: the theory of general relativity. There are whole books written on the subject of general relativity, explain-ing how he derived it, what the implications were and what it means. For the purposes of this chapter, we will look at the bare essentials only. Partly because we don't need all the mathematical detail of the theory, and partly because I barely understand the equations myself.

The first thing we need to appreciate is just how many dimensions our universe has. We can describe the shape and size of an object using three measurements – height, depth and breadth. But imagine an object that existed in all three of these dimensions but didn't exist for any amount of time. If you don't exist for any amount of time then you don't exist at all. Time, therefore, has to be thought of as a dimension in its own right.

Our universe has what physicists call '3+1' dimensions. We say it like this, rather than simply saying '4', because it signifies that the time dimension is a little different to the other three – although they are closely linked and how you move through the space dimensions has an impact on how you move through the time one.

During the 1910s, physicists were becoming aware of this linkage between the four dimensions and began referring to them as two aspects of a single background fabric to the Universe called 'spacetime' (a term coined by one of Einstein's heroes, Hermann Minkowski).

The key proposition of general relativity is that this spacetime mate-rial gets distorted by the presence of mass or energy. Einstein referred to spacetime, rather oddly, as a 'mollusc' because he imagined it as a jelly-like substance that warped and wobbled in the presence of objects.[2]

In empty parts of the Universe the background spacetime is smooth and undisturbed, but near a large mass spacetime becomes bent and curved.

The analogy that often gets used is to imagine a sheet of rubber pulled tightly across a room and a heavy ball placed in the middle. Since this ball has mass, it distorts the sheet and anything trying to move in a straight line from one end to the other will be inclined to follow the curves of the sheet, drifting inadvertently towards the ball in the middle.

It's a helpful mental picture and certainly easier to visualise than drawing curvature in 3D, but, for the sake of it, spacetime curvature looks something like this:

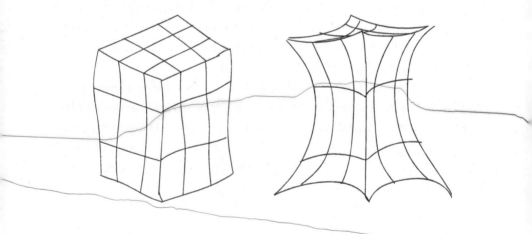

The image on the left shows the coordinates of empty spacetime without energy or mass inside. When we put an object into the centre of that region, however, it curves everything towards it, making it look more like the image on the right. The top of that warped cube is akin to the rubber sheet in the other analogy. If you try to move in a straight line through a region of curved spacetime, you will end up drifting inwards, towards the mass at the centre.

Passing through curved spacetime also changes the nature of any measurements you might take. Imagine setting off in a rocket with enough fuel to travel 10 kilometres (6.2 miles). If you happened to pass near a planet you would be moving through curved spacetime and would suddenly find your fuel only got you 9 kilometres (5.6 miles) because the spatial dimensions have been distorted around you. To make it past the planet you would need more fuel than if you were travelling through 'flat' spacetime.

This curvature of spacetime causes objects with mass to attract each other, which we refer to as the 'force' of gravity between them. Gravity is, in essence, a by-product of curvature in spacetime. The larger the object, the more spacetime curvature and thus the stronger the gravitational field it generates.

Clock measurements also distort near large masses because when space curves, so does time. This doesn't affect you on a day-to-day basis but technically your head is aging faster than your feet because it's slightly further away from the Earth's mass. That also means the core of the Earth is about two and a half years younger than its crust, despite having formed at the same time, because the spacetime is more heavily distorted inside the planet than on its surface.

The mathematics of general relativity are fiendishly complex (see Appendix III) but the outcomes are straightforward and can be summarised succinctly. Spacetime is distorted by energy/mass, and an object with energy/mass gets influenced by spacetime. As the physicist John Archibald Wheeler put it: 'mass tells spacetime how to curve and space-time tells mass how to move'.[3]

BENDING LIGHT

Once Einstein had figured out the equations for general relativity he had to put his hypothesis to the test. For this, he turned to a specific prediction his equations made which could be observed with a decent telescope – the gravitational lensing effect.

Since anything with mass curves the spacetime around it, a beam of light will bend as it goes past, following the curved tracks of empty space. The larger the object, the more the beam of light will be curved and, if the object is heavy enough, we might actually be able to see what is behind it.

In the diagram below, the trajectory of light from the twinkly star wants to continue on up the page, but because the fabric of spacetime through which it's travelling is being curved by that large and perfectly drawn planet, it ends up veering off course, getting flung towards our eye.

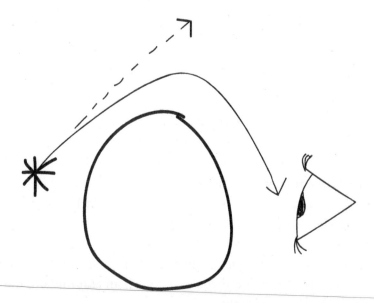

In order to confirm this prediction, Einstein persuaded his friend Erwin Freundlich to travel all the way to Crimea where he could study an impending solar eclipse. During the eclipse it would be possible to point a telescope directly at the Sun's edge for up to two minutes and see if light from the stars behind it were deflected by the curving of spacetime.

Freundlich agreed but unfortunately for him (and really the whole of Europe), three weeks prior to his expedition, Archduke Franz Ferdinand of Austria was assassinated by a terrorist group, triggering the First World War. Since Crimea and Germany were on opposite sides, Freundlich suddenly found himself behind enemy lines and was detained for a whole month by soldiers believing him to be a spy carrying 'mysterious equipment' in his luggage. It was a telescope.[4]

Einstein was disgruntled by this and had to wait for another solar eclipse to occur on 29 May 1919, this time observable from West Africa. The English astronomer Arthur Eddington got permission from his government to avoid military service in order to take a team to test general relativity, and took a series of measurements that validated Einstein's hypothesis.

General relativity was now a fully fledged theory, which could be used to describe anything from a ball rolling down a ramp to something as complex as the expansion of the Universe itself. Which is actually what Georges Lemaître used it for.

HERE WE GO AGAIN

Red-shift discoveries had shown the Universe to be expanding in every direction but what if, Lemaître speculated, it was the other way around. Just like al-Sijzi and his astrolabe had shown it could be Earth moving and stars remaining still, Lemaître wondered if instead of empty space sitting still and galaxies moving through it, it could be galaxies sitting still and the fabric of spacetime expanding between them.

Lemaître was, I guess for the fun of it, seeing if it was feasible to create a hypothetical universe in which spacetime was stretching and, in the process, discovered that if you made such an assumption, the solution matched what we observe.

Furthermore, if we played everything in reverse then at some point in the distant past, all of reality would shrink down to a point where spacetime would knot into a ball and the spacetime curvature would become so compact we would have no way of describing it. Lemaître called this early state of the Universe 'the primeval atom' but nowadays it is referred to as the cosmic singularity.[5]

A singularity in physics jargon refers to a set of conditions where our knowledge no longer works. It's a fancy way of saying 'we haven't got a clue what's going on here'.

In Lemaître's hypothesis, all matter, all energy, all space and all time were previously compacted into an infinitely dense, infinitely hot, infinitesimally small point that, for some reason, began to expand. Clearly it was in need of some snazzy gut-adjustable shirts. If only someone would invent such a thing.

Lemaître did not know where this singularity came from or what made it start expanding in the first place. He just concluded that as the

Universe began to expand, spacetime unfurled, matter cooled and things started behaving in a way we call 'existing'.

Pope Pius XII was so excited at this that he publicly praised Lemaître's hypothesis and said it proved the Universe had a beginning, meaning there had to be a god. Lemaître was not so comfortable with this, however, and requested an audience with the Pope to dissuade him from taking that position.[6]

Lemaître was a priest but he did not want to use science to prove or disprove the existence of God. For one thing, he felt there should be a clear distinction between matters of theology and matters of science, and for another, he was not sure his hypothesis made any such claim. His solutions to the general relativity equations just pointed to an earliest point we could recognise as 'the familiar Universe'. It made no comment whatsoever about what things were like inside the singularity that came before.

THE MOST MISLEADING NAME IN SCIENCE

One of Lemaître's most vocal opponents was the astronomer Fred Hoyle, who favoured an alternative explanation for galactic red-shift: the steady-state hypothesis.

In the steady-state view, the Universe has always been the same kind of thing and spacetime has always had the same characteristics. Instead of spacetime expanding, new matter is constantly being created to fill in the gaps as things move apart. The overall density of the Universe remains constant and thus we never have to handle the difficulties of a singularity.

During a radio interview on 28 March 1948, Hoyle explained that he thought Lemaître's expanding singularity was 'about as elegant as a party girl jumping out of a cake', and at one point ridiculed the idea that 'all the matter in the Universe was created in one big bang'.[7] The name immediately stuck in everyone's mind, despite being notoriously misleading.

For starters, the singularity wasn't big. It was an infinitesimally small dot (that's the point). Furthermore, it didn't go bang because there were no sound waves and thus the stretching process would have happened in complete silence. The 'big bang' name tends to conjure images of a tiny lump of matter blasting itself apart and filling the surrounding emptiness with explosion, but this is selling the whole thing short. It's much weirder.

It wasn't matter blasting outwards to fill empty space so much as empty space itself coming into existence, untangling and taking on the characteristics of 3+1 dimensions. It also wasn't a long-ago event. It's happening right now. The Universe is currently 'big banging' and we're inside it. The 'singularity expansion' would be a much better name, but big bang caught on, underselling the idea and making it sound silly – much to Hoyle's delight.

Initially, both the big-bang and steady-state hypotheses were mathematically equivalent and there was no way of deciding who was right. But then in 1948 two cosmologists named Ralph Alpher and Robert Herman hit on something that could act as a litmus test to distinguish between them.[8]

According to the big-bang hypothesis, as spacetime expanded the Universe cooled down and free-floating particles joined up to form atoms. This process of cooling would involve a significant loss of energy, released from particles in the form of short wavelength gamma light.

Over the eons, as spacetime stretched, this light would get stretched too, becoming long wavelength microwaves by the present day. The entire Universe should therefore have a cosmic microwave background (called the CMB for short) humming throughout it and such a feature was not part of the steady-state hypothesis. Searching for the CMB would therefore settle whose hypothesis was right.

You probably know how it turned out. The longest running sitcom in television history isn't called *The Steady State Theory*.

GATHERING ROUND THE HORN

In 1964 two physicists named Robert Wilson and Arno Penzias were taking measurements with the Holmdel Horn antenna in New Jersey, a 6-metre contraption built to pick up radio and microwave light from the stars. Their intention was to use the Horn for radio spectroscopy of the

Milky Way but when they began their experiments, they detected an incessant hiss in the background of their signals.

Penzias went out to the antenna to see what was causing the problem and discovered that two pigeons had made a home for themselves inside the Horn and, being pigeons, had liberally deposited what he tactfully described as a white material over everything. It's good to know that at some point in the history of figuring out the Universe, a scientist had sit down and think, 'Now, what's the scientific term for pigeon poop?'

The pigeons were caged and sent away so the research could continue. However, it turned out they were homing pigeons and after reaching their destination, flew right back to the Horn. At this point, one of the researchers (they have never confessed which one it was) took control of the situation in the simplest way possible and killed the birds with a shotgun.[9]

When they went back to their scanner, however, they found that the microwave disturbance was still there. Not only that, they were picking it up in every direction of the sky, all the time. The entire Universe was apparently humming with microwave energy and at this point they started to get confused (as well as hopefully a little bit guilty because those pigeons didn't have to die).

It was only after phoning their cosmologist friend down the road, Bob Dicke, that the truth emerged. They had accidentally stumbled upon the cosmic microwave background without even looking for it. The big-bang hypothesis now had evidence in its favour and thus it too was upgraded to the rank of 'theory'.

The CMB has become one of the most important tools cosmologists have for studying the early stages of the Universe because it is a microwave photograph of what was going on right after the expansion started. Really, when we look at the CMB, we are seeing what the plasma of the big-bang furnace looked like in its very final moment. You can even detect it yourself with an old CRT television. If you tune it to empty static, what you are seeing is the scrambled mesh of microwave signals travelling through space and about 1 per cent of that static is from the CMB; the afterglow of the big bang.

For their discovery, Penzias and Wilson were awarded the Nobel Prize for Physics, although some complained that because they made their discovery by accident and weren't even sure what they were looking at, the prize was undeserved. I see the logic in that position, but considering what happened to the last two living things who got in their way, I'm going to stay quiet.

PART II
EVERYTHING WE DON'T KNOW ABOUT SPACE

CHAPTER FOUR

Big Problems for the Big Bang

The Dark Triad

It's a vital part of scientific method to question well-accepted theories and find their limits. The big bang is accepted as a theory because it has robust evidence in its favour, but it is by no means the end of the story and it doesn't explain everything.

To be a scientist means you never rest on your laurels and as soon as you have an explanation for something you start probing for weaknesses. The big bang theory currently faces three awkward issues . . .

1 The horizon problem.
2 What's outside the Universe?
3 What caused the big bang?

Like a Balloon

The cosmic microwave background is the oldest thing in the Universe we can actually see, because any earlier than 380,000 years and space itself wasn't transparent. Everything was a hot, glowing soup of particles and it would be like trying to see the inside of a fire. Saying it's the oldest thing we can see is also saying it is the most distant thing because in cosmology older equals further away.

It takes time for a beam of light to reach us, so when we look at something far away we are seeing it as it was when the light left, not how it is at the time we are looking. Even seeing the Moon in real time is impossible because it is one 'light second' away, i.e. it takes a full second for moonlight to reach Earth.

Technically this is true for anything you observe in your daily life, it's just that the travel time is so short it becomes negligible and we act as if we are seeing the world around us as it happens, when in fact we are seeing everything with a brief time delay.

Studying the CMB is a way of analysing the earliest era possible, but there's a problem when we do so. The CMB contains within it an imprint of the temperature fluctuations taking place at the time it formed, and we have found that it's the same temperature in every direction, even on opposite sides of the Universe. Why is that a problem? Because 380,000 years isn't long enough for the Universe to have cooled down to an even temperature.

In the earliest moments after expansion, the temperature of everything would have been fluctuating wildly, with some places being much hotter than others. If you leave a bunch of hot and cold things next to each other they will converge on an average temperature, but it takes time for this to happen because energy has to move from the hot regions to the cold ones.

In the case of the Universe this time would have been much longer than 380,000 years because at that point in time the heat energy would not have finished travelling from one side to the other. We say that opposite ends of space were 'over the horizon' from each other's perspective, hence the name: horizon problem.

At the point the CMB formed there should have been a lot of temperature differences still going on, so how come everything had settled to the same average temperature if there hadn't been enough time?

In the diagram below, the vertical line represents the point at which the CMB formed and the star represents where the big bang expansion *did* begin 13.8 billion years ago. The problem with the temperature having settled by that point is that it would have needed the Universe to start much earlier, at the point indicated by the circle and dotted lines.

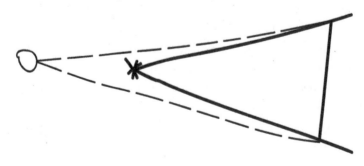

The most widely discussed, although far from accepted, answer to this puzzle is called the 'inflation hypothesis', proposed by Alan Guth in 1979. According to Guth, rather than expanding smoothly like in the cone above, something much uglier happened at the start of the big bang.

In the first moment after expansion, a trillionth of a trillionth of a trillionth of a second, the Universe was the same temperature everywhere because it was too small for fluctuations to fit inside it. If you boil a saucepan of water the surface will jump about with bubbles and dips, but if you boil a thimble of water, it's too small for fluctuations to form and everything stays the same.

If the Universe suddenly underwent some sort of mega-expansion at this moment when everything was the same, different sides of the Universe would end up looking identical because there would not have been time for a temperature fluctuation to even occur. The expansion of the Universe would end up looking more like an elongated bell, as in the diagram below (the vertical line still showing formation of the CMB) rather than a smooth cone.

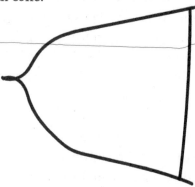

Even though opposite sides of space would be over the horizon from each other by the time the CMB formed, when inflation happened they would have been intimately close. After this mysterious inflation, the expansion slowed down to its normal speed and the CMB formed, recording the same background temperature everywhere.

Inflation is by no means an accepted theory and it generates all sorts of unanswered questions of its own, the biggest being: what caused inflation and why did it stop?

In March 2014, there was a brief flurry of excitement when a team of researchers operating the BICEP2 telescope at the South Pole thought they had found traces of the inflationary period recorded in the CMB.

Unfortunately, after the rest of the scientific community started looking at the data, it turned out to be a false start. What the research team had actually detected and mistook for inflationary disturbances in the CMB was space dust.[1]

THE EDGE OF FOREVER

Every second the Universe's diameter grows by many millions of metres, expanding three times faster than the speed of light. Although nothing goes faster than light within the confines of spacetime, spacetime itself is allowed to stretch at whatever rate it pleases. Defining the 'edge' of the Universe can therefore mean different things depending on your perspective.

Let's start by revisiting Hubble's discovery that more distant galaxies are moving away from us faster. The further a galaxy is, the faster it is moving away, so eventually there must be a galaxy so far out that its speed away from us would be equal to the speed of its light towards us. Light moving in our direction from such a galaxy would seemingly never reach us, making it forever invisible.

This limit to what we should be able to see is called the Hubble horizon and it exists 14.4 billion light years away. That number might seem a little surprising because, if the Universe is only 13.8 billion years old, surely nothing could be more than 13.8 billion light years away? But remember that although nothing *inside* could have travelled that far, empty spacetime itself can go faster and galaxies are not really moving themselves, just the spacetime between them.

The next Universal 'edge' cosmologists talk about is the furthest distance we could get to if we travelled away from Earth at the speed of light. Obviously, a spacecraft travelling at such a speed is impossible but if it were, we would find that it couldn't go forever.

If we sent a light-speed rocket towards a distant galaxy, spacetime would be stretching around us, moving that galaxy further and further away. The fastest we could chase after it would be light speed, but because spacetime can stretch faster than that, we would never quite catch up. It would be like running up an escalator that is moving down faster than our legs can move, preventing us from reaching the top. We call this point that we could reach the cosmic event horizon and it sits 16 billion light years away.

Then there is another 'edge' to consider – by far the strangest of the three. Any galaxies beyond the Hubble horizon are moving away from us faster than the speed of light. It would be reasonable therefore to assume that they are invisible to Earth because they are moving away faster than their beams can reach us. But because spacetime is expanding, the size of our Hubble horizon is expanding too. This means our Hubble horizon will eventually reach these approaching beams of light, which were previously outside what we could see, and engulf them, meaning they will eventually move slowly towards us.

There is still a limit to how far we can observe, of course, because the Hubble horizon hasn't caught up with everything and we call this distance the particle horizon – the *actual* farthest distance we can detect. The particle horizon is 46 billion light years away, which means the observable Universe has a current diameter of 93 billion light years and counting. That's bigger than at least two football pitches.

THE SHAPE OF THINGS TO COME

The particle horizon of the observable Universe represents the edge of knowledge, and that edge is expanding. For something to expand suggests it has some sort of shape, though. What kind of shape is the Universe?

There are three possible answers allowed by general relativity, all with different rules, and we call these possibly allowed Universe shapes 'spaces', appropriately enough.

The simplest 'space' is that the Universe is simply expanding into more emptiness. In this view, the rules of geometry work nicely and nothing unusual happens when we get to the edge of space – there's just more space. We call this option Minkowski space (after Hermann Minkowski, the guy who coined the term spacetime).

Unfortunately the other two spaces are impossible to visualise or draw because they are shapes which exist in 4+1 dimensions rather than the 3+1 we are used to.

Our brains cannot picture a higher spatial dimension, but there is no reason to insist our universe *must* be spatially 3D. We have the terms left/right, up/down and forward/back to denote movement along three spatial axes, but there could be a fourth axis we just aren't seeing. Theoretical physicists sometimes use the words 'ana' and 'kata' to describe moving along this axis even if we cannot imagine what it would look like. So, let's talk about this dimension in more detail.

In the 1884 satirical novel *Flatland* by Edwin Abbott the characters are all 2D shapes living on the surface of a 2+1D Universe. One day the hero (a square) is visited by a creature who can seemingly appear out of nothing, expand to the shape of a circle, then shrink away and vanish. The stranger explains to the square that there is nothing to fear because he is in fact a 3D 'sphere' and can slice his way into 2+1D Flatland, being viewed in cross section by its inhabitants.

The 2D square is then taken on a journey to visit a 1D universe where all the inhabitants live on a line, and then to a zero-dimensional universe inhabited by a single-point being, the god of all. The book ends with the 2D square asking if there are fourth and fifth spatial dimensions, at which point the sphere scoffs and tells the square not to be ridiculous.

The Flatland Universe is a 2+1D Minkowski space. The angles of a triangle add up to 180 degrees, parallel lines never meet and so on. But supposing we rolled the surface of Flatland into a sphere. From a Flatlander's perspective, if they travelled in a straight line they would eventually return to the point where they set off. They would also find that two parallel lines would meet eventually and triangle angles would add up to more than 180 degrees.

The problem for them is that they live on the 2D surface of a 3D object and are too small to see the curvature – the same way we are too small to see the curvature of our own planet but know that setting off in one direction eventually brings us back to where we set off. The Flatlanders could still deduce that higher dimensions existed, however, if they observed how light behaved over distances great enough to be bent by the curve.

We call this kind of inwardly curving shape a 'de Sitter space' after the Dutch mathematician Willem de Sitter, and our universe could conceivably be a higher dimensional version of that. In this space our 3D universe is considered the 'surface' of a 4D hyper-sphere and if that hyper-sphere increases its size so will its surface, i.e. our 3+1D universe. There would be no apparent edge from the perspective of *our* spacetime but there would be from the perspective of any 4+1D beings looking 'kata' at us from outside the hypersphere.

If our universe exists in de Sitter space, it would not be infinite and if we set off in a straight line we would eventually get back to where we started. Furthermore, if two beams of light travelled in parallel lines for long enough they would eventually meet, following the higher dimensional curve of the spacetime. We cannot picture such a space, obviously, but equations have no difficulty describing how a 4+1D universe would behave, and it's pretty much the same as ours.

The third possibility, probably the strangest, is that we could bend Flatland into a horse-saddle shape. This is called an anti de Sitter space and, much like flat Minkowski space, it's infinite.

An anti de Sitter space would look a bit like living inside a magnifying glass to its inhabitants. When you peer at something through such a lens, the bit in the centre looks to be the right shape but everything towards the edges appears squashed and distorted around the rim.

In the same way, living in an anti de Sitter space would make the very distant edges of spacetime appear squashed, until you headed off to meet them. When you did that, the distorted image would suddenly adjust to take on the right proportions (like moving the magnifying glass to a different place) and where you were previously standing would now look distorted if you looked back over your shoulder.

Here's a handy table to summarise the possible spaces our universe could take.

Spacetime is . . .	Flatland is on the 2D surface of a . . .	Our universe is on the 3D surface of a . . .	Features
Bent Inwards (de Sitter)	Sphere	Hypersphere	Finite Get back where you started Parallel lines meet
Flat (Minkowski)	Sheet	N/A	Infinite Parallel lines remain parallel
Bent Outwards (anti de Sitter)	Saddle	Hyper-saddle	Infinite Parallel lines diverge

At the moment we do not know which kind of space our universe adopts. It looks like a Minkowski space from the long-distance measurements we have taken so far (which is useful because that would be the easiest to handle) but we have not done enough high-precision experiments to know for sure.

WHERE DID IT ALL COME FROM?

Whichever shape the Universe is, the question still remains: what exists in these higher spatial dimensions our universe expands into?

I'm afraid the shortest, simplest and most honest answer is not satisfying at all . . . we just don't know. The cosmic singularity was so unlike anything we can describe (by definition of being a singularity it is beyond understanding) so we cannot say much about what is outside, if that's even the right word to use.

Just to confuse things even more, because time is an integral part of the Universe's dimensions, if we talk about a place where the Universe isn't, we are *also* talking about a place where time might not exist. Going outside the Universe or going further back than the singularity might be impossible because nothing exists outside space or before time.

For most of human history we have had to answer these questions by writing poetic myths about them. In the ancient Greek poet Orpheus's story of creation (composition date unknown), Chronos, the god of time, lays an egg from which all the other gods and the Universe spring. Orpheus neatly avoids the awkward question of where Chronos/time himself came from.

In the Chinese philosophical text *Huainanzi* (before 139 BCE) by Lui An, there is only a featureless void to begin with. It is a nebulous environment where up, down, forward, back, left and right are all synonymous and from this contradictory nothingness, qi, the essential life force, is born, giving rise to the rest of reality. Once again, the writer bypasses the details of specifically how this happened.

Theoretical physics runs into the same problem: 13.8 billion years ago the big-bang expansion started and spacetime took form but, before that, spacetime was not there, which means time did not exist. In fact, the sentences I just typed are not strictly allowed. To say the Universe 'started' is to say there was time passing. But there wasn't. I shouldn't even have used the words 'was' or 'wasn't' just now because they imply the singularity was sitting at a certain point in time. It wasn't. Arggghhh!

It's not just our equations that break down when we get to a singularity; our language does too. Spacetime is not relevant for a singularity so we cannot sensibly talk about anything happening. There is no before, no after, no during, no cause, no effect and no words to describe such a state of affairs.

Asking where the Universe came from or what made it happen is like asking which letter comes before A in the alphabet. You either have to invent a new letter, conclude there is no letter before A or argue that it somehow loops back to Z. All we can say with confidence is that there is a universe and there was an earliest point we can talk about. To go any further is no longer the realm of science.

Although, perhaps going outside science might be inevitable in this arena. After all, science is the study of the natural world, and the question of how the natural world started might best be handled with a supernatural answer. Maybe these questions are just as much about philosophy as physics.

Stephen Hawking argued that since there was no such thing as 'before' the singularity, the law of cause and effect does not apply and therefore we do not need to explain what 'caused' the Universe: it didn't need a cause. On the other side of the debate, people such as the philosopher and Christian theologian William Lane Craig argue that if time began with the Universe there must have been a transcendent cause for time itself – what he believes to be evidence for a creator god.

I will leave it to the reader to decide if they think a supernatural creator is a good or bad explanation for the Universe. Science neither proves nor disproves the existence of one. All I will say is that the question 'Why is there something rather than nothing?' is arguably the most profound and important question we can ask. We ought to pursue all avenues and not dismiss anything without being very cautious. Dismissing or accepting the idea of a god, without taking our time to really think about it, would be extremely unwise.

CHAPTER FIVE

Help, Most of Our Universe Is Missing!

Unknown Elements

There is a lot about the Universe we know, but even more that we don't. Quite literally. Most of the Universe is either missing or completely hidden from us and the stuff we *can* see accounts for a fraction of what's really out there.

One of the most glaring problems facing modern astrophysicists is called the 'cosmological lithium problem' and goes as follows: If the big-bang theory is correct then the first atoms to form would have been the smallest ones possible: hydrogen, deuterium, helium-3 and helium-4. We can calculate what ought to be the abundance of each type by the present day and, when we go looking, we find the numbers are not only close to our predictions, they are bang on.

In fact, to get this level of match between theory and data by chance, it would be like placing four dartboards at random throughout the whole of China and dropping four darts from a plane, striking each board dead in the bullseye. Technically it could be a coincidence that our darts landed exactly where the boards were placed, but it seems far more likely that our calculations are correct.

However, there is a huge discrepancy between prediction and observation when it comes to the amount of lithium-7, the next heaviest atom. Our current big-bang theory predicts there should be three times as much lithium-7 as we actually observe and, at the time of writing, nobody can explain where it has gone.

This doesn't mean the big-bang theory needs to be thrown out – the rest of the evidence supports it astonishingly precisely – but the lithium problem is an important reminder that we still have a lot to work out.

LOPSIDED

Another problem facing our current understanding of the Universe is the curious fact that it is made of matter. I know what you're thinking: *Gee, Tim, you don't say . . . what else should the Universe be made of: unicorn tears and fairy mucus?*

Thing is, if we follow the laws of particle physics, there should be twice as much Universe as there is or, from a different perspective, no Universe whatsoever.

Let's consider electrons – the particles that orbit atoms, one of the most abundant particles found in nature. Electrons have a negative electric charge but positively charged electrons can be made under extreme circumstances such as nuclear explosions or radioactive decay. We call negative electrons 'regular matter' and the positively charged variation 'antimatter'.

It isn't just electrons that have antimatter twins either. All the particles in your body, electrons, protons, neutrons, etc., are regular matter particles but have potential antimatter counterparts, which we can synthesise in a lab. What doesn't make sense is that regular matter is extremely common while antimatter is rare.

The main reason antimatter doesn't last very long is that antimatter is attracted to regular matter and, when they meet, they annihilate each other, converting all their energy into neutral particles of light called photons. (NB: you're allowed to talk about light being made of either waves or particles, pretty much depending on what mood you're in. Physics is just like that.)

Preserving antimatter is tricky, therefore, because as soon as it comes anywhere near regular matter (which most of the world is made of) the electrons zip towards each other and everything vanishes in a flash.

The reverse can happen as well. Light beams can spontaneously split in half, turning into an electron and anti-electron (i.e. the positively charged electron, sometimes called a positron), which quickly draw back towards each other and re-convert back to light. The diagram below shows a light beam splitting as it moves across the page, before the generated electron and anti-electron boomerang together and return to normal light.

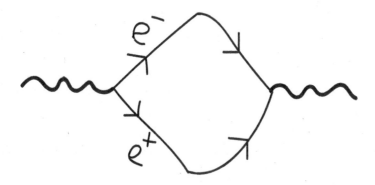

Now, here is the rub. In the early Universe, just after expansion started, everything was made of light. In order for that light to turn into normal particles, it had to simultaneously generate antiparticles as well . . . which would have recombined with all that regular matter.

A universe made exclusively from matter shouldn't be possible because there should have been an anti-universe generated at the same time to cancel it out. The Universe should be made of light, and yet it isn't because somehow things became lopsided and all the antimatter disappeared. Where did it go? We have absolutely no idea.

You're Moving Too Fast

Galaxies spin for the same reason moons orbit planets and planets orbit suns. Any two objects with mass will be drawn towards each other, entering a dynamic arrangement in which the more massive object sits in the middle and the lighter object goes round the outside (although technically the heavier object doesn't sit perfectly still, it gyrates a bit).

Einstein's equations for general relativity describe how these orbit systems look and the rule is simple: the heavier something is, the faster it has to move in order to remain in orbit. If a planet slows down, the gravitational pull from its sun will pull the two together and the planet will be gobbled up.

The same can be said of galaxies. The heavier the galaxy is, the faster it has to rotate to avoid being sucked into the black hole at the centre (see next chapter), like a mouse paddling desperately around a draining plug-hole. To put it simply: heavy galaxies spin fast and light galaxies spin slow.

But in 1933 the Swiss astronomer Fritz Zwicky was studying the rotation of galactic clusters and made a strange discovery. Most galaxies rotate too fast for the amount of stars they contain, as if they were heavier than they appeared. Zwicky wondered if galaxies contained a substance that does not give out light, making it hard to detect, and called this hypothetical substance, creatively, dark matter.

Zwicky's dark matter was taken seriously in academic circles but there were lots of alternative explanations for what was going on. Perhaps there were more nebulae in these galaxies than assumed. Or perhaps it was as simple as there being lots of heavy planets around each sun. Nothing to lose sleep over.

Until, in 1976, the astronomer Vera Rubin made a game-changing discovery of her own. Galaxies are not just spinning too fast, they are spinning much, much, much, much, much too fast. Extra planets and nebulae could explain a minor discrepancy with general relativity but Rubin found that most galaxies rotate six to eight times faster than should be possible. This wasn't a minor calculation error, this was something big.

Following Rubin's measurements, we were left with two explanations for the overly fast rotation of galaxies. Either we had Einstein's equations wrong or there really is a mysterious material filling the Universe, lurking just outside our detection methods.

Maybe Einstein Got It Wrong?

This is the position some physicists have taken in the dark-matter debate – that Einstein's equations are mistaken. It's a perfectly respectable position. Claiming the Universe is full of a new type of substance is a bold statement so the simplest response is to say, 'No, it probably isn't.'

Then again, suggesting general relativity is incorrect is also a pretty incendiary thing to say. As a theory, it has been put to the test many times and has always yielded solid results, not only with things such as gravitational lensing but in the formation of galaxies, the cosmic microwave background, the way planets orbit suns, the fact that GPS satellites work, and the prediction of black holes and gravitational waves (see next chapter).

It's also not the first time general relativity has been under attack. For decades, the amount of regular (not anti or dark) matter in the Universe seemed to be half what it should have been. General relativity predicted a certain amount of protons and neutrons – collectively called baryons – but it was twice what we could actually see.

But then in 2018 the missing baryons were discovered and there was exactly the amount general relativity predicted.[1] They turned out to be hiding in enormous bridges, hundreds of light years long, connecting galaxies together like a cosmic internet. We had never seen them because we had never really thought to look at the spaces between galaxies rather than inside them. General relativity is not above question, of course, but every time we have challenged the theory, the theory has won.

Another important discovery in favour of dark matter is the so-called Bullet Cluster. Out in the direction of the Carina constellation, we have recently discovered something truly gargantuan – a collision between two galactic clusters. Not two galaxies. Two whole *groups* of galaxies. The official designation for them is 1E 0657-558 but 'Bullet Cluster' sounds cooler.

As these two titanic objects meld in mid-space, they spin around a common centre of mass, the same way planets and suns do. But the interesting thing about the Bullet Cluster is that the centre of mass they orbit is nowhere near where it should be.

Calculating the centre of mass based on stars alone should give a location vastly displaced from where it actually is and the only explanation which seems to fit is that there is additional matter hidden in these galaxies, throwing off their orbits.

To explain the Bullet Cluster without dark matter would require us to overhaul not just general relativity but the very idea of mass and gravity itself. In all honesty there is no reason we shouldn't be forced to do this

– no theory is above question – but the majority of astronomers have come down in favour of Zwicky and Rubin. Dark matter is assumed to be real.

So . . . What Is Dark Matter?

Frankly, there are only two things we know about dark matter. 1) It's matter. 2) It's dark.

Its reluctance to interact with visible light means the normal techniques we might use to observe it are out the window, so we can only make judgements about it based on the one property we know it has: mass.

There are many types of particle we currently know about. There are electrons, neutrinos, muons, tauons, quarks, photons, gluons, weak bosons, Higgses, and they all have a variety of properties with names such as charge, mass, magnetic field, spin and so on.

What's really surprising, however, is that all the particles we have discovered thus far have something in common: they are very, very light. Especially photons. They are literally light. (This is the part where my accompanying drummer does that snare-drum punchline thing.)

The lightest particles we know of are photons and gluons, which are both massless. Then come neutrinos, followed by electrons and so on, right up to the heaviest known particle: the top quark.

We usually express particle masses in units called eVs (which stands for 'electron volts', although in this context it doesn't have anything to do with electricity) because kilograms are so big that the numbers look silly. For what it's worth, though, the top quark has a mass of 0.00000000000000000000000031 kg, and that's as heavy as they come.

The heaviest a particle can hypothetically get, however, is a different story. The laws of physics permit particles to exist with any mass up to 0.02 mg, called the Planck mass, after the quantum physicist Max Planck. In the same way the Universe has a speed limit (light speed), particles have a mass limit.

The Planck mass is about equal to a grain of sand and, although that seems unimpressive from our perspective as big humans, if we compare it to how heavy all the other known particles are, it's many orders of magnitude heavier.

Looking at it this way, it seems unlikely we have discovered every type of particle the Universe has to offer. It's probably a safe bet that somewhere on that scale there are particles that do not interact with light but have significant mass. If there are lots of these particles floating about, we would expect to find them huddled into galaxies, making them appear heavier than they ought to be.

The Large Hydron Collider (LHC) particle detector at CERN has been searching for dark-matter particles since its inauguration in 2009 but, at the time of writing, has found nothing. In fact, the LHC might not be big enough to agitate these particles into showing themselves (as a general rule, the heavier a particle is the more energy it takes to detect it).

This is one of the reasons CERN wants to build an even bigger particle detector with a diameter of 100 kilometres (62 miles), four times the size of the LHC. The current working title for this monstrous machine is the FCC, which stands for Future Circular Collider, and so dark matter's days of hiding may be numbered. It might be thinking to itself:

> So the FCC won't let me be,
> Or let me be me,
> So let me see.
> They tried to detect me with the LHC,
> But things feel so empty without me.[2]

(My sincere apologies to Eminem . . . and to the legal team who had to investigate if I was allowed to use Eminem lyrics for the sake of a particle physics joke.)

Big Fat Blunder

When Einstein penned his equations for General Relativity there was a sticking point. They originally predicted that the Universe should be shrinking. Gravity pulls on matter so everything should be pulling together into a ball, making the Universe impossible. Evidence suggests the Universe is *not* impossible, so Einstein had to modify his theory to match.

He decided to add an 'anti-gravity' force, which he called the cosmological constant, to his equations. Although never detected, it was assumed to be counteracting gravity in order to keep the Universe in balance.

When it was explained to him that the Universe was not collapsing because it was already expanding (either from the steady state or the big bang) he scrapped the cosmological constant and, according to his friend George Gamow, referred to it as his biggest blunder.[3] Personally, I'd say his biggest blunder was not using his fame to promote the sale of girdle-shirts. But then again, he was the genius, not me.

Anti-gravity was forgotten about until 1998 when two groups of physicists in California, led independently by Saul Perlmutter and Adam Riess, discovered that certain supernovae behave in a way which might make it not such a blunder after all.[4]

Supernovae are what happens when a heavy star dies under extreme circumstances. We have already seen that small stars shrivel up, re-pressurise their cores and slowly expand before drifting apart to form a nebula (see Chaper One). A supernova is a similar process dialled up to eleven.

The outer layers of a massive star shrink in the same way, but at such a velocity that they bounce off the inner core and explode back outwards, tearing the star to pieces in an explosion so bright it radiates the same amount of energy in a few minutes that our sun produces over its 10-billion-year lifetime.

There are a few ways in which this can happen. Type I supernovae are the result of one star leaching matter from another nearby. As it hoovers up plasma from its neighbour, it becomes heavier until its own weight can no longer be supported by the core and a collapse is triggered. Type II supernovae are the result of stars simply being overweight, so that when the core runs out of fuel it sucks the outer surface in catastrophically.

Supernovae are well-understood phenomena and we have been observing them for a long time. The earliest record dates back to the fifth-century Chinese historian Fan Ye who tells of a mysterious event that took place in 185 CE, when astronomers in China saw a new star appearing between the constellations of Centaurus and Circinus. This 'guest star' was so bright it could be seen in daylight and remained in the sky for eight months. We now know that they were witnessing the supernova event SN 185 exploding 86 trillion kilometres (52 trillion miles) away.

Supernovae always occur in the same fashion and emit the same amount of light, which means we can use their luminosity as a reference to figure out how far away things are. By analysing their red-shift we can also calculate how fast they are moving, and it turns out that once again things are moving too fast. Not just galactic rotation this time: everything.

By looking at the red-shift and brightness of supernovae near and distant to us, Perlmutter and Riess were able to measure how fast the Universe is currently expanding and compare it to how fast it was expanding in the past. What they found was that older supernovae were not flying apart as fast as they are today. Which means the Universe is not only expanding, it's somehow getting faster.

If a mass is changing velocity, we say a force is acting on it, and something with the ability to exert a force is said to have energy. Given the amount of energy in the Universe from the big bang, we can calculate what the speed of expansion ought to be, so if things are getting faster, something with energy must be responsible.

It was given the insanely cool name 'dark energy' but it has be stressed that it's a completely separate phenomenon to dark matter. Dark matter isn't too difficult to come up with an explanation for – it's some kind of particle we haven't found. But dark energy? That's a much trickier topic.

It might seem like a minor curiosity not worth fussing over, if it weren't for a rather embarrassing fact. Because Einstein's $E = mc^2$ equation tells us energy and mass are interchangeable (in some sense they are the same), we can calculate how much mass a given amount of energy has. Calculations vary slightly from research team to research team but the mass of dark energy makes up roughly 75 per cent of the Universe. Dark matter then makes up 23 per cent, and all the stuff we can actually see – every galaxy, nebula, baryon bridge, solar system, star, comet, asteroid and planet – makes up 2 per cent of the Universe.

So . . . What Is Dark Energy?

Dark energy is harder to explain than dark matter. We know it's spread throughout the Universe but it is a very dilute substance. In fact, dark energy is so dilute it went more or less unnoticed for the first 9 billion years of the Universal timeline.

During those years, the momentum of the big bang was slowing down due to gravity, but around 5 billion years ago (late August on the cosmic calendar) galaxies were spread out so much that gravity was too weak to hold them together.

At this point, gravity's pull was briefly equal to dark energy's push and, for a moment, the Universe neither slowed nor accelerated. But dark energy has an advantage: it's the same strength everywhere, while gravity diminishes with distance. Once we reached the point where the two were equal, gravity could no longer counteract the effect of dark energy and the Universe began to speed up.

Some have suggested that dark energy is a new force of nature along with gravity, electromagnetism and the nuclear forces. Some have said it could be a new feature of spacetime itself and that Einstein's cosmological constant needs to be revisited. Some have speculated it could be the result of quantum fields all around us and that the energy of empty space is a lot higher than previously assumed. Nobody knows what it is or, disconcertingly, what it means for the future.

Where Are We Headed?

At the moment we don't know enough about dark energy or its interaction with gravity to say what the future has in store for us. Here are the four most widely discussed and debated ideas.

- The Big Freeze. Let's say dark energy and gravity stay in their current relationship and the Universe keeps expanding. In that scenario, all the galaxies will drift apart over trillions of years until light can no longer move between them. The night skies will go black and all the energy in the Universe will slowly disperse, spreading to form a stable, boring, nothingness.
- The Big Crunch. If dark energy's current reign is only a temporary effect then the Universe will eventually stop accelerating. If you pull on an elastic band you can increase the rate of stretch at first, but eventually you get to the elastic limit and it will pull back. If gravity starts to win again, spacetime will reverse its current behaviour and everything will fly towards a reverse big bang, getting smushed into another singularity.
- The Big Bounce. Similar to the big crunch only this time the Universe will reach its singularity and repeat the whole thing over. The big bang will happen once more and existence will cycle back out, replaying itself in an infinite chain of expansion and contraction. Perhaps each repeated cycle is identical or perhaps each bounce is slightly different, giving each universe its own unique timeline.
- The Big Rip. Let's say dark energy is just getting started. We don't know what it is so perhaps as time increases, so will its strength. Maybe the outward push on the Universe will not only overtake gravity but become too much for spacetime to handle, allowing the forces of dark energy to tear the cosmos apart, destroying existence itself. So y'know, bad news for reality. Good news for anyone with an overdue library book.

CHAPTER SIX

The Heart of Darkness

Faintest Ripples

On 11 February 2016, international headlines were made when astronomers announced the discovery of gravitational waves, something predicted by general relativity one hundred years previously.

Consider two stars orbiting each other. As they spiral around one another they create a repeating rhythm of fluctuations in spacetime itself – ripples made of gravity.

These gravitational waves should be detectable, but they would be extremely weak because spacetime is not disturbed easily. It's not a substance like water, which will ripple under the slightest provocation. Spacetime is more like thick treacle and even really heavy objects will only produce tiny waves when they move. In order to detect these microscopic vibrations, we needed serious hardware. Cue LIGO – the Laser Interferometer Gravitational-Wave Observatory.

The design of the LIGO is as elegant as it is simple. First, you take a laser beam and direct it at a partially silvered mirror. If you get the angle just right, you can split the beam in two; half of it bouncing off at a right angle and the other half carrying on through the glass, the way a window can reflect and be transparent at the same time.

The diagram here shows the laser beam approaching from the left, hitting the mirror and half bouncing upward/half continuing on its journey.

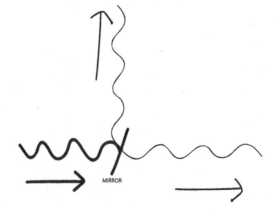

MIRROR

Next, we put two regular mirrors in the path of both beams, so that they bounce back along their original course, still in perfect lockstep with each other.

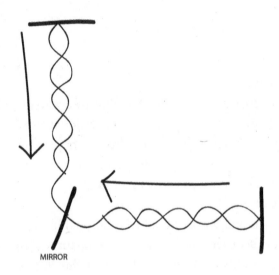

MIRROR

When the beams re-join at the silvered mirror, they will do exactly what they did before. Half of each beam will go through the glass while half will bounce off at 90 degrees.

If we stick a detector at the bottom of the diagram, we will apparently see an ordinary beam arriving. There would be no way of telling that half that beam had gone on a vertical journey and half had gone on a horizontal one.

But what if something about the horizontal pathway changed? What if the distance from the slanted mirror to the reflecting one were squashed? If we moved the horizontal mirror a bit closer, the horizontal beam would no longer be travelling the same distance as the vertical beam, meaning that when they recombined, they would be out of phase with each other and the detector would pick up two beams interfering, rather than one coherent beam.

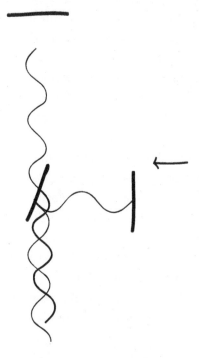

This is precisely what a gravitational wave would do: change the distance between the mirrors as it alters the dimensions of space itself. That means if a gravitational wave passed through the detector, we would get minute changes in the laser-beam path lengths, leading to interference.

The LIGO project currently consists of three such setups located at Hanford in Washington, Livingston in Louisiana and Santo Stefano a Macerata in Italy (the Italian one is called VIRGO, but all three stations collaborate and share data). The arms of the Italian detector are 2 kilometres (1.2 miles) long, with the American detectors being twice that size.

The first gravitational waves LIGO detected passed through the Earth on 14 September 2015, shifting the path length of the lasers by only a few nanometres.[1] Since then, a dozen more have been detected, winning Nobel Prizes for Kip Thorne, Barry Barish and Rainer Weiss, LIGO's chief designers.

But what was causing these ripples? Planets and suns produce gravitational waves as they orbit each other, but they would be far too weak to detect. In order to generate even the faintest gravitational wave, the objects spinning around each other would have to be colossal. The only thing that could account for these gravitational waves would have to be astronomy's most shadowy mystery: black holes.

Dark Star

It is sometimes said that the first person to propose the existence of black holes was the English scientist and priest John Michell in 1783.[2] Every object, be it asteroid, planet or star, has an escape velocity – the speed you have to be moving at in order to escape its gravitational pull without additional energy.

For Earth, this velocity is 11,200 metres per second, i.e. if you threw a ball directly upward at this speed it would leave orbit, assuming it didn't catch fire as it tore through the atmosphere.

This doesn't mean, however, that anything travelling slower than escape velocity will never reach orbit. Rockets being launched are typically going much slower than this speed, but because they are constantly burning fuel they are adding extra propulsion energy to their flight. Escape velocity is really the velocity you need to be travelling for *unassisted* escape and the more massive the planet, the higher that velocity needs to be.

Jupiter's escape velocity is 60,000 m/s and the Sun's is 617,000. What John Michell pondered was the existence of a star so massive its escape velocity would become 300,000,000 m/s: the cosmic speed limit.

Even a beam of light would not be able to escape a star with an escape velocity this large so if you stood on the surface of such a star (don't, by the way) and pointed a torch upward the beam would curve back and return towards you.

Michell called these hypothetical objects 'dark stars' because their emitted light always gets pulled back inwards, making them undetectable to outside telescopes. The idea is superb and inventive, but the modern theory of black holes is a whole lot weirder.

Kicking Mass

When a Type II supernova happens (that's the one where the star is so massive its outer layers bounce off its inner core) what is left behind is an ultra-dense spinning ball of material called a neutron star.

The exact composition of a neutron star is a topic of debate but it's often assumed to be a material in which the electrons which are normally orbiting atoms have been squeezed onto their protons, forming a substance composed of pure neutron. This material would technically be an element but it belongs outside the periodic table and is sometimes nicknamed 'neutronium'.

A neutron star is like an atomic nucleus the size of a mountain, spinning several hundred times a second, generating a whopping magnetic field on each rotation. These neutron stars were first detected in 1967 by Jocelyn Bell and Antony Hewish, who named them pulsars because they pulse electromagnetic energy as they spin.

One teaspoon of neutronium would weigh 10 million tons and the escape velocity of the neutron star would be about 40 per cent the speed of light. We often say the spacetime curvature around a neutron star is extremely 'steep' in reference to how steep the sides of the pit would be in the classic rubber-sheet/bowling-ball analogy. Although a neutron star would really be more like a wrecking ball.

The planet PSR J1719-1438b is believed to orbit a neutron star rather than an ordinary one, but this is a rare occurrence as almost anything approaching a neutron star is liable to get pulled into its gravitational field and crushed.[3] Including other neutron stars.

In October 2018, LIGO detected a collision between two neutron stars – a phenomenon termed a kilonova. Kilonovae do not give off as much light as supernovae so they're harder to see, but it is suspected that the pressure of such an event is responsible for forming the heavy elements of the periodic table from gold to uranium. If you own any gold jewellery you may actually own a tiny piece of kilonova. And if you own any uranium . . . why do you?

It has also been proposed that really large neutron stars could work up enough gravity to collapse themselves further, so that the neutrons themselves break down to the fundamental particles of matter, quarks. These hypothetical 'quark stars' have never been observed, although a few of the neutron stars we have detected (with names such as XTE J1739-285) seem to be too dense for what we would expect, making them good quark-star candidates.[4] And there is no reason we have to stop at neutron stars or quark stars.

In 1916 the German physicist Karl Schwarzschild was playing around with Einstein's equations (you know, if there's nothing on the telly or whatever) when he decided to see what happened if you pushed the mass of a collapsed star higher and higher, making its gravitational pull stronger and stronger.

The heavier things got in his calculations, the smaller the radius of the star became, giving it greater and greater density. The result was a chain reaction of gravity, which grew until what was once a bunch of particles transformed on the page into something unrecognisable.

If the gravity of a dead star were great enough, the spacetime around it would become so distorted the dimensions of space and time would swap places. Time would become a spatial dimension and would point in a direction, i.e. the past would exist behind you physically (not just behind you metaphorically) and the future would exist literally in front of you.

In such an extreme circumstance, time would force you towards the dead star in the centre because the future lies there and trying to go backwards in space would require you to go backwards in time as well.

This is another way of saying you have no choice but to move inwards. It's not just that the escape velocity is greater than light speed, spacetime itself becomes so messed up you can no longer get out because 'out' has become synonymous with 'the past'. Your future exists along one path now . . . towards the centre.

If we return to the analogy of the rubber sheet, what has happened in this instance is that the mass on the sheet is so great, it has pulled the gravity well of spacetime into a vertical tube.

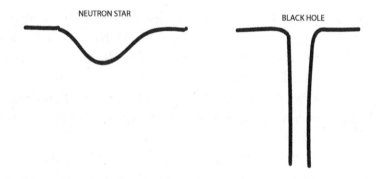

NEUTRON STAR

BLACK HOLE

This is impossible to draw in 3D, but the outcome can be said simply: if you approach this object you get to a point from which there is no return, what physicists call the 'event horizon'. The event horizon would be analogous to the tipping point on the diagram above, but in 3D it is more like a bubble surrounding the extreme spacetime curvature. We call the resulting structure a black hole.

Schwarzschild also showed that anything could theoretically be crushed to the point of forming its own event horizon bubble. For planet Earth it would happen if you crushed it to 9 millimetres ($\frac{1}{3}$ inch) across and for the Sun, it would be 3 kilometres (1.9 miles).

Fortunately, there is no way to do this to the Earth or Sun and only enormous stars above what is called the Chandrasekhar limit (after the Indian astrophysicist Subrahmanyan Chandrasekhar) will actually form black holes.

For decades, astronomers argued over whether Schwarzschild's solutions were real things or just quirky predictions of general relativity, but the discovery of gravitational waves has settled the debate. The only way of working up enough spacetime distortion to create gravitational waves such as the ones we picked up would be if two black holes collided – an event so extreme we do not even have a name for it. I propose the term: Schwarzonova.

HOLLYWOOD MAGIC

On 10 April 2019 the Event Horizon Telescope, a network of eight radio-telescopes positioned around the globe, turned the whole Earth into a giant antenna aimed at the galaxy M87, 520 quintillion kilometres (323 million miles) away, in the hope of taking a photograph of its central black hole.

Using software written by a team of researchers led by twenty-nine-year-old grad student Katie Bouman, the Event Horizon recorded 5 peta-bytes of data (that's about 5000 years' worth of MP3s) taking up half a ton of hard drives. In fact, the data recorded from the Event Horizon Telescope was so big it could not be sent over the internet and had to be transported manually by airplane in order to reach Bouman's lab, where it was stitched together.[5]

The image obtained is an orange/yellow ring, looking like a glowing doughnut in space. What's truly exciting about the black-hole photo-graph is not just that it proves beyond all shadow of a doubt that black holes are real, it matches the predictions made by general relativity about what they should look like.

Before we captured this photograph we had to rely on computer simu-lations. The most accurate one ever created was featured at the climax of the Christopher Nolan sci-fi movie *Interstellar* in which the hero, played by Matthew McConaughey, approaches the fictional black hole, Gargantua.[6]

It looks like something Christopher Nolan dreamed up to make a jaw-dropping finale, but the black hole in *Interstellar* was designed by a team of physicists led by Kip Thorne, one of the Nobel-Prize winning scien-tists who built the LIGO. The computer-generated effects took 30 people to design and used 800 terabytes of data, with each frame taking 100 hours to process.

Thorne originally drafted the story for *Interstellar* as an informal sequel to the Robert Zemeckis movie *Contact* starring Jodie Foster and, coincidentally, Matthew McConaughey. His idea was picked up by Steven Spielberg but eventually got co-scripted and directed by Nolan, while Thorne served as a scientific advisor.[7]

The special effects computers used for the movie were programmed to incorporate general relativity into their design as well as Karl

Schwarzschild's solutions to the equations. The simulation ended up being so scientifically reliable, in fact, that Thorne even published the results in the journal *Classical and Quantum Gravity*.[8]

Although some critics nit-picked about *Interstellar*'s occasional fudging of physics to make things look cool, overall it does a fantastic job of conveying what these things are like. Besides, the only other major film to feature a black hole as central to its plot is the Disney movie *The Black Hole* in which the interior of the event horizon is inhabited by satanic robots. I'll take Matthew McConaughey any day.

OUTSIDE IN

What would we see if we were to get close enough to a black hole to explore it, or even fall in? Much of the answer depends on the type of black hole we're considering. Black holes come in two sizes: stellar and supermassive. Stellar black holes have a mass five to ten times that of our sun and are created via supernovae the way we have already seen. These are the most common types of black hole, with potentially millions flitting about the galaxy.

Supermassive black holes, on the other hand, are quite rare and we don't honestly know how they form. They are the beasts that sit at the centres of galaxies and probably aid in their formation. While moons go round planets and planets go round suns, the suns in a galaxy orbit a supermassive black hole in the middle. The Milky Way's supermassive black hole is called Sagittarius A* and has a mass 2.6 million times that of our sun.

They also come in two varieties – stationary (called Schwarzschild black holes) or rotating (called Kerr black holes, named after mathematician Roy Kerr). That doesn't mean the event horizon of a Kerr black hole is a literal surface that is rotating, mind you. It's the spacetime itself that is swirling, a bit like a whirlpool.

An important misconception worth addressing at this point is that black holes are not like vacuum cleaners sucking stuff up. In the same way the Moon isn't sucked into the Earth, you can orbit a black hole, provided you move at a fast enough speed. Black holes are just regions of intense gravity and while you can fall in, you don't get sucked in. Which is just as well because the black hole A0620-00, the closest to Earth, is only 3000 light years away and would have swallowed us by now.

If you were to approach a stellar black hole you probably wouldn't see anything special. Just a bunch of stuff orbiting an apparently empty part of space. A supermassive black hole, on the other hand, would be a lot more impressive because all the dust and matter orbiting it gets heated up as it grinds together, forming a glowing ring, which is what we saw in Katie Bouman and her team's black-hole photograph from 2019.

Keep going inwards and you'll reach a point called the photon sphere, where beams of light are bent into a circle by the spacetime curvature. If it's a supermassive black hole you're approaching, you might also notice powerful streams of particles, called relativistic jets, spitting out of the top and bottom. Sometimes these point directly at us; when we first detected them we named them quasars (quasi-stellar objects), but we don't know how they form.

The most widely suspected hypothesis is that particles going round the event horizon of a black hole are often charged, which generates a magnetic field, much the same way a loop of wire carrying a current gives rise to magnetism. Black holes could thus have magnetic fields, which channel particles to the north and south poles, ejecting them at close to the speed of light.

Then, inside this layer of photons, you get to the actual black hole itself. Which you can't see. Obviously. It's black. Although the reason they are black is not strictly because light gets sucked into them . . . stay tuned.

THE ULTIMATE BOUNDARY

John Michell's idea of a dark star pulling light towards it is easy to visualise, but in reality black holes are more bizarre. A better visualisation of the event horizon is to think of it like the point on a waterfall where the water starts toppling over the edge and you can no longer swim against the current.

The physicist Paul Painlevé (who was also Prime Minister of France on two occasions) realised that the 'water' in this 'waterfall' is space itself, i.e. empty space is moving towards the middle of a black hole faster than the speed of light. A beam of light shone away from the centre of a black hole will still end up moving inwards because 'away from the centre' does not exist. All directions point inwards once you cross the event horizon. And the shape of space is not the only thing that gets all whacky. Because space and time are linked, a black hole distorts time just as much as space, leading to some pretty paradoxical results.

Anyone watching you fall towards a black hole would see your clock running slower compared to theirs because you would be experiencing time at a stretched rate. If they kept watching, your image would drag and drag until it eventually froze at the point time became infinitely slow, i.e. your image would stop moving altogether.

I say 'eventually' as if this is something your friend could genuinely observe, but really it takes an infinite amount of time for your image to reach a full stop. Someone watching you from the outside would never quite see you fall through the event horizon. They would just see your movements getting slower and slower as you got closer and closer to the edge. Forever.

Your image would also get dimmer because the light waves coming from your body get stretched from visible to infrared, to microwave, to radio and so on. Or, if you want to think of light in particle terms (like I said, you're allowed to do both), the time between each particle being sent towards an observer gets longer, making your image dimmer.

From the perspective of the outside Universe, nothing is ever seen going into a black hole, just super close to the edge, and this is the real reason black holes are black. It's because from an outsider's perspective black holes don't appear to contain anything. Not even the original star that formed them.

When a star collapses during supernova the 'point of no return bubble' appears (from the Universe's perspective) to originate at the centre of the star and move outwards, smearing the star across its surface, even though we know logically it is the other way around.

The blackness of a black hole is the result of it appearing empty – a place where nothing exists. We know this is an illusion, of course, but a black hole gives the appearance of being a perfect vacuum: an eventless region of the Universe, hence the name 'event horizon'.

Our friend observing us entering the black hole would never actually see us going past the boundary, but from our perspective we would sail right through undisturbed. If we were to look backwards at the point of crossing the event horizon, however, we would see the outside Universe speeding up like a video on fast-forward – the opposite of what the Universe sees for us.

We would be able to watch the entire future history of time zipping before our eyes, although annoyingly the image would gradually shrink to a point so we wouldn't be able to see anything useful.

Staying on the edge of an event horizon isn't easy, of course, so in practice observing the future history of the Universe wouldn't be straightforward. But, if we could somehow stay on the event horizon and eject ourselves back out a few moments after we got stuck there, we would find ourselves emerging in the distant future. Black holes are genuine one-way time machines.

Onwards and Inwards to Nothing

The closer we get to a black hole's centre, the less we understand the physics going on. Our facts run out, our equations become shaky and we have to replace confidence with speculation. The interior of a black hole is a 3D funnel made of moving spacetime, ushering us inescapably towards a nothingness, so figuring out what happens in this place is not easy.

It's also not as if we could send a probe into a black hole because the information could never be transmitted back. Even attaching a cable to the probe wouldn't work because the very electrons inside the wire would get stuck on the inner side of the event horizon.

We are fairly confident, however, that the interior of a black hole is not actually dark. They are black from the outside because nothing appears to be going on inside, but to someone inside the bubble everything would be illuminated. If we looked back at the place we fell in, we would see the outside Universe as a glowing point of light showering towards us and we might even be able to send signals to other people nearby who have also fallen in.

As we get closer to the centre, though, we would start to notice the force of gravity increasing via what's called a 'squared relationship'. That means that if you halve the distance between you and the centre, gravity quadruples. Halve it again and gravity becomes sixteen times stronger so the force on your feet (I'm assuming you dived in feet first for some reason) becomes much greater than the force on your head.

The overall effect is that while all of you is being pulled in, your feet are nearer the centre so every millimetre closer increases the gravity by a factor of a thousand, stretching you in a process physicists call (genuinely) 'spaghettification'.

For supermassive black holes it could be a long time, possibly even years, before this starts to happen, but for a stellar black hole spaghettification can happen before you even cross the event horizon. We can even watch it happening right now to one of the stars in the Arp 299 galaxy, which is in the process of being spaghettified by a nearby stellar black hole.

It sounds gruesome but the good news is you probably wouldn't feel any pain because although your feet are being stretched, the signal of 'ouch' would not reach your brain as the nerves become elongated as well.

Then, finally, once you have been stretched into a thread the width of a proton, you would reach the nucleus of the spacetime waterfall and all our theories turn into question marks. The heart of a black hole is where gravity becomes so extreme that space and time no longer exist and, once again, we are talking about a singularity.

It isn't right to call this black-hole singularity a 'thing', however, because it isn't one. The centre of a black hole is a place with no describable properties. It isn't an object, it doesn't have a size and it doesn't exist in time.

The only thing we know about singularities for sure is that if the black hole is rotating, the singularity region gets stretched into a ring (which should obviously be called a ringularity) but, other than that, we know nothing. The singularity is a hole in spacetime. But if it is a hole, we need to ask: what's on the other side?

THE SPACETIME SUBWAY

I remember once discovering I had torn a hole in my trousers. In a moment of astonishing stupidity, I began turning them inside-out in the hope of getting rid of the hole (it had been a long day). Obviously I was confronted with the other side of the hole, along with a feeling of deep shame and a conviction that I should never become a science educator. I mean, what exactly was my plan even if it had worked? Walk around for the rest of the day with my trousers inside out?

If you cut a hole in a sheet of fabric it is automatically a hole in the opposite side of the sheet, and black holes, at least mathematically, are no different. Karl Schwarzschild's equations describe two kinds of singularity existing together. One which pulls everything in and one which pushes everything out. On paper it looks as though something reaching a black-hole singularity can emerge in a reverse process, which we have named a white hole.

A white hole is a theoretically described region that forces spacetime outwards, i.e. there is an event-horizon boundary you cannot enter, since the curvature of spacetime is too steep in the opposite direction.

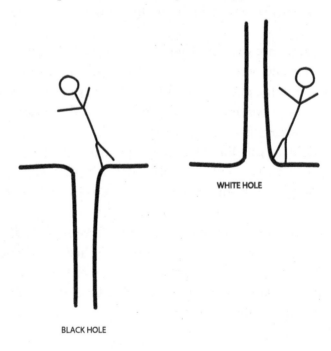

WHITE HOLE

BLACK HOLE

However, we need to be careful here. Just because an equation predicts something doesn't mean it has to exist. For example, say you wanted to put some fencing along the side of your lawn. You know your lawn has an area of 9 square metres, so you square-root that value to get the length of each side. The square roots of nine are +3 and −3 but you wouldn't go to your local gardening store and ask for 'minus three metres of fencing'. You recognise that the equation is made up of symbols that don't know what they are describing.

Schwarzschild's equations contain both the existence of black holes and their negatives, but a lot of physicists are sceptical that 'negative black hole' means anything. White holes have never been observed and there is a case for saying they never could, because they violate other laws of the universe.

To get around that problem, the physicist Freeman Dyson suggested that if white holes do exist they might have to do so in a different universe altogether.

If we return to our bedsheet analogy and imagine poking a hole through the fabric, that hole will exist on the other side of the sheet. A tiny bug living on the surface of this universe could go through the hole and find itself in an upside-down reality.

Ordinarily, to get to this 'reversiverse', the bug would have to travel across the sheet, curve over the hem (whatever that would mean) and navigate along the underside. A rip in spacetime, however, would allow the bug to make the journey to another universe in an instant.

The connecting pathway between a white hole and a black hole is called an Einstein–Rosen bridge, named after Einstein and his collaborator Nathan Rosen who developed the theory describing them. They are usually known by the more colourful name 'wormhole', invented by John Archibald Wheeler, who thought of them as a tube bored through an apple, allowing a worm to shortcut from one side to the other without having to navigate the surface.[9]

The idea was then proposed (by Einstein himself) that if we forget the black-hole/white-hole openings it might be possible to have just the tube on its own. A wormhole like that might even be able to connect two points within the same universe, allowing us to travel to different regions of space that would normally take us centuries to reach.

These are called 'traversable wormholes' while the black-hole/white-hole one-way systems between parallel universes are called 'Schwarzschild wormholes'.

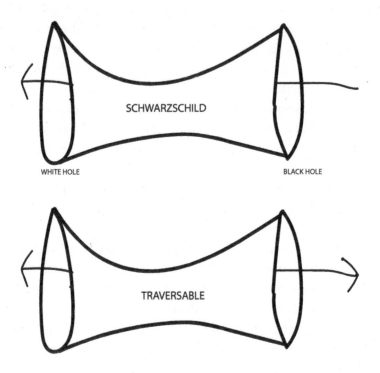

SCHWARZSCHILD

WHITE HOLE   BLACK HOLE

TRAVERSABLE

Because all of this is based on theory and calculation rather than hard fact, wormholes are a topic of thrilling disagreement. Some physicists argue that you could never get to the other end of a traversable wormhole because spacetime narrows in the middle (or collapses entirely) and you would become trapped forever. Others have suggested that the right kind of particles could prop a wormhole open, allowing us to ride them like a transit system to wherever and whenever the two ends existed.

Nobody knows what we might find on the other side of a black-hole singularity, if you even go anywhere at all. Maybe you just get destroyed. Or what if every black-hole singularity leads to a white-hole singularity, which creates a universe of its own? Perhaps the big bang we live inside is a white hole created by another universe? Perhaps there is a universe squashed inside every black hole, giving the impression to creatures living inside that they are taking up space in their dimensions, while taking up none in ours.

A black-hole singularity is a pure mystery so you can pretty much invent whatever you want to go inside it. And if the idea of an entire universe inside every black hole sounds fantastical, just wait until we take a closer look at the event horizon.

CHAPTER SEVEN

Holograms, Loops and Strings

The Nexus of Knowledge

There is a major theoretical reason black holes are difficult to work with (not to mention it's hard to stop people falling into them in the lab). Currently there are two sets of physical laws we use to describe the Universe. There's the physics of heavy stuff, which is handled by general relativity, and there's the physics of small stuff, which is handled by quantum mechanics.

The singularity of a black hole is unique because it is both very heavy and very small (taking up zero dimensions), meaning you need to combine general relativity and quantum mechanics to make sense of it. This is the point where we stumble, because so far nobody has been able to come up with a theory that includes both quantum mechanics and general relativity.

Studying black holes is worth doing, though, because they provide an insight into how both sets of physical laws might combine. Understanding their characteristics might therefore put us on the path to an ultimate Theory of Everything. So far, the only person to successfully join any part of quantum mechanics with general relativity nearly broke physics in the process.

Hairy Holes

The event horizon of a black hole is a smooth region of spacetime with nothing coming out. The only features a black hole can have are mass, rotation and electric charge. The physicist Jacob Bekenstein playfully remarked that black holes are so featureless they 'have no hair' and once something goes in its properties never re-emerge. This has always been a key assumption.

Another key assumption is that the more things fall into a black hole the heavier it becomes, meaning black holes should get bigger over time. But in 1974, a young English physicist battling motor neurone disease changed both these assumptions.

Stephen Hawking's life story is an inspiration to anyone who hears it, but he was not only celebrated because of his courageous battle with paralysis. He was additionally one of the most brilliant theoretical physicists of the twentieth century.

In the 1 March 1974 issue of *Nature*, Hawking published an article titled 'Black hole explosions?', through which he revealed that he had found a way to combine some of general relativity with quantum mechanics. In this landmark paper, Hawking showed that when you try to combine both theories you end up with an inescapable conclusion: not only does some stuff escape the event horizon, a black hole eventually shrinks into nothingness. The reason for this is that black holes have another property along with the conventional three: temperature.

We know it's impossible for an object to have a temperature of absolute zero (−273 °C) because at absolute zero particles would become fully stationary. This is impossible due to something in quantum mechanics called the Heisenberg uncertainty principle, which says, among other things, that every particle in the universe jitters about so no particle can ever be truly motionless.

This principle also applies to the background quantum fields of empty space, out of which particles are formed. Empty space is not really motionless; it's bouncing with energy that usually vanishes too quickly for us to detect. What Hawking did was show that this energy jittering happens on the surface of an event horizon, leading to the startling conclusion that black holes sometimes radiate energy into space. Or, to put it another way: they have a temperature.

This is happening slowly (black holes are frightfully cold) but a tiny amount of energy leaks away from them and if a black hole loses energy it is in the process of shrinking. Given a few quadrillion years, even a supermassive black hole will shrivel up to nothingness.

We call this energy being given off by a black hole 'Hawking radiation' in his honour, but while we know it must be there, we don't know what causes it. This isn't unheard of in physics history. Isaac Newton had no idea what started the planets orbiting but he knew something had done so, so he left that bit blank for later scientists to figure out. Hawking radiation is quite similar.

The proposed explanations for how Hawking radiation might arise rely on quantum mechanical phenomena that would require a major detour, so I'll put the details into Appendix IV for the curious. The important takeaway is that black holes are not as bald as we thought – they have a bit of stubble.

Hawking's paper made this energy loss inevitable but, more importantly than any equation, there is experimental evidence to prove he was right. We obviously can't build a real black hole because we can't control light that well, but we can control sound so it's possible to create a sound-wave version of a black hole and study it to get clues about how the real ones work.

By manipulating vibrations in an ultra-cold fluid, it's possible to set up a region inside a container where the speed of a flowing liquid is faster than sound vibrations trying to move against it: a sonic event horizon.

They are called 'dumb holes' (because they trap sound rather than light) and in May 2019 Jeff Steinhauer reported that dumb holes his team had created occasionally spat out unexplained micro-ripples in the liquid moving *away* from the event horizon.[1] They appear to be the sound-wave equivalent of Hawking radiation.

I should mention at this point that there is an apocryphal story, related by Alan Lightman, that the equally legendary physicist Richard Feynman came to the same conclusion as Hawking and filled a blackboard with equations proving black holes emitted radiation.

According to the story, Feynman left his equations overnight but by the time anyone realised their significance the cleaner had scrubbed it all away and the equations were gone forever, much like the black holes themselves.[2] It's a fun story but it has to be taken with a pinch of salt because I don't know any cleaner foolish enough to wipe a physicist's blackboard clean.

WE WANT INFORMATION

One of the key principles of physics, so basic we take it for granted, is that cause and effect is true (inside the Universe at least). We can describe any event by linking it to what happened the moment before or extrapolating forwards to predict what will happen the moment after.

The echoes of every conversation you have ever had are preserved in the surface atoms of the walls of your house and, if you had a slightly different conversation at any point in the past, the atoms would be vibrating slightly differently today. Practically, it isn't possible to keep track of all this information but in theory if you know the present state of something you can wind backwards to figure out what used to be happening.

Physicists measure the link between past and present states using a property they call 'information'. It has a strict mathematical definition but the idea is that as time goes from past to present to future the amount of information must stay the same. A particle's current state can be backtracked to a previous one and so on, otherwise cause and effect would not work. But if Hawking's evaporating black hole idea is right, we have a serious problem.

When particles cross an event horizon they take their information with them. This information is stored inside the black hole and if we were to wind backwards through time, we would see the particles come flying back out with their information intact.

But now let's factor in the black-hole evaporation. As a black hole loses energy the event horizon will shrink until it closes in on the singularity and the whole thing disappears. The black hole is now gone along with all the information it once contained. Where there once was a black hole there is now empty space with Hawking radiation flying away from it.

We can analyse this Hawking radiation and extrapolate back to figure out how the event horizon (the outside) of the black hole was behaving, but Hawking radiation tells us nothing about what was *inside* the black hole. Hawking radiation forms on the edge/outside of the event horizon and it's cut off from the interior. Whatever goes on inside cannot get back out, so the information stored inside the black hole has just been deleted from existence.

This means if we look at a region of empty space and conclude there used to be a black hole there, we can never find out what went in. Bearing in mind our whole understanding of past, present and future is based on the idea that information never disappears, evaporating black holes throw a bit of a spanner in the works. Hawking radiation seems to imply that information *can* disappear and whole chunks of reality can get erased without us knowing, leading to what is called the 'information paradox'.

Hawking was pretty confident about this discovery and even entered into a bet with the physicist John Preskill that nobody would be able to find a way out. For decades, physicists tried to rescue the laws of physics and, as you can probably guess by now, the answer leads to some startling paradoxes of its own.

THE UNIVERSE IS SAVED

If we go back to the point at which a particle falls over the event horizon, we recall that an image of that particle remains frozen on the outside due to time dilation, i.e. we never actually witness anything fall in.

That means the information is effectively 'imprinted' on the event horizon so when Hawking radiation is generated, the two could theoretically influence each other. Hawking radiation coming away from a black hole will behave differently depending on what is imprinted on the event horizon and therefore the information might not be destroyed. It could be transferred from the event horizon to the Hawking radiation, and back to the rest of the Universe.

The mechanism of how this might happen is unclear because we don't know how Hawking radiation is formed in the first place, but the physicists Gerard 't Hooft and Leonard Susskind showed that it was possible, at least in mathematical theory, for information to be preserved by event-horizon fluctuations.

Hawking conceded the bet with John Preskill and agreed that there was a way to save information after all. His forfeit was to buy Preskill a baseball encyclopaedia, although he joked that it would have made sense to burn the encyclopaedia and send Preskill the ashes to represent how scrambled its information had become.[3]

That all sounds neat, with the information paradox getting wrapped up in a pretty bow. But not so fast. The amount of information in the Universe cannot change, we know that much. But if the information is stored on the event horizon *and* falls into the black hole, we have effectively photocopied it. Information apparently doubles at the point of crossing the horizon, which is just as bad as it being deleted. Either the information goes in or it doesn't, right?

The solution to this new paradox was developed by Juan Maldacena and Leonard Susskind, who showed that there was a potential way to avoid copying information, provided we did something unprecedented. In essence, they proposed that there already are two sets of information: one is the 3D particle itself and the other is the 2D surface around it. This is gonna get a little weird . . .

THE UNIVERSE IS SAVED, BUT IT MIGHT BE A HOLOGRAM

Imagine a hypothetical cube made entirely of electrons. Such an object would be hard to make, but humour me for a moment. The six walls of this cube are made of negatively charged particles skittering around the surface. Now imagine firing another electron into the box from the outside. Because that electron is also negatively charged, when it pierces the wall of the box it will change the positions of all the surface electrons. Those disturbances in the surface are measurable and if we look closely we could deduce all the information we wanted to about the box's interior.

In fact, we never need to study the internal electron directly because we can observe how the surface of the box behaves. We could calculate how long ago the electron entered, how fast it was going, where it entered and even deduce its current location inside.

It would be like analysing the ripples of a swimming pool to figure out how long ago someone jumped in and where they were swimming under the water. In practice, it would not be easy, but in theory the 2D outer layer would tell us everything we needed to know about the 3D interior.

Maldacena and Susskind showed that this principle of working out higher-dimensional information from lower dimensional surroundings could be applied to quantum information. If this is true, then the information of any 3D space might be subtly encoded in the 2D boundary encasing it.

When an object falls into a black hole, the 3D information (the particle) goes in while the 2D information (the particle's surroundings) stays outside, later to be carried away by the Hawking radiation. I can't think of a way to sum this up in a neat little diagram, so instead here is a sketch I made of an onion, which will probably confuse anyone skimming through the book.

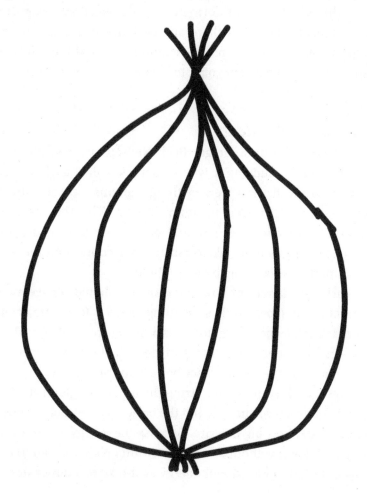

Maldacena and Susskind's idea is called the 'holographic principle' and says that higher dimensional information is really unnecessary. In the same way that you can create the illusion of a 3D image on a 2D surface – a hologram – you can work out everything about a region of spacetime from its surface area. A 3D object is a kind of projection from a 2D surface.

But remember how we might be living inside a black hole – that old chestnut? Does that mean our 3D Universe might be a hologram from some 2D surface on the outside of reality itself? The answer, disturbingly, is maybe.

But how did they arrive at this whacky conclusion? What made them suggest higher dimensional information could be stored lower dimensionally in the first place? Is there any evidence to support the holographic principle or is it just pretty equations to avoid the information paradox? To answer that we're going to have to learn some string theory.

THE UNIVERSE IS STRINGS?

In 1970 Leonard Susskind, Holger Bech Nielsen and Yoichiro Nambu were independently studying a type of particle called a gluon. In the usual quantum mechanical view of things, gluons zip back and forth between the quarks that make up atomic nuclei. What they were beginning to wonder was whether it would be simpler to consider gluons shooting to and fro as more like elastic strings connecting the quarks, rather than particles travelling between them.

Their early calculations of how gluon strings should behave matched experiments nicely (specifically something called a Regge trajectory, in which the rotational speed of gluon–quark structures matches what you would expect for strings rather than particles). So they began speculating what other particles they could describe as being string-like.

One important feature of this 'string theory' is that gluon strings can take all sorts of different energies but the basic way they behave does not change, no matter how big they are. Because the energy of gluon strings can go up and down without affecting anything else, you can make the equations easier by talking about energy as if it were a dimension in its own right. If you move a particle to and fro along the three spatial axes, it doesn't affect anything, so we might as well do the same with the energy axis and imagine that gluon strings can metaphorically 'move along the energy direction' too.

Three physicists named Tamiaki Yoneya, Joel Scherk and John H. Schwarz independently carried out this calculation and discovered that strings vibrating in a dimension of energy no longer look like any particles known to quantum mechanics.[4] In fact, they looked like a hypothetical particle called a graviton. Gravitons had been talked about for years as a possible way to reconcile quantum mechanics with gravity – the idea being that all objects emit and absorb graviton particles between them causing attraction – but nobody had ever taken them seriously.

This strange new string theory took a while to get going and there were lots of false starts, but gradually people began to wonder if the physicists might be on to something. It was a theory that included both quantum particles and gravity – something no other theory offered.

In the string-theory view, particles are essentially illusions. Instead of little chunks of matter there is only one kind of substance in the Universe: strands of energy which vibrate in different ways. A string can vibrate one way and take on the properties of an electron, then vibrate another way and take on the properties of a gluon. It is only from our big-person perspective, unable to see the vibrations, that we mistake them for point-like particles.

You might want to ask at this point what the strings are made of, but the answer is, rather unhelpfully, more string. The smallest kinds of string are called F-strings (for fundamental) and they can melt together to make what are called D-strings (named for the German mathematician Johann Dirichlet). Strings can also be stretched into 2D surfaces called branes (short for membranes) and branes can be smushed on top of each other to create 3D structures called bulks.

Whereas quantum mechanics has hundreds of particles with different properties, in string theory there is only one thing (string) and you only need to know its length, which dictates the mass. Once you know that, you can work out all the possible vibrations it can have, which gives you a list of all the remaining properties of the particle. In other words, knowing the mass would predict the charge, the magnetic field and everything else about it.

However, to account for all the particles we know of, 3+1 dimensions are clearly not enough. Once you have vibrated your string vertically, horizontally and lengthwise, that's all you can do, which only accounts for three kinds of particle. In order to account for all the known particles, we need other directions for our strings to vibrate in and, subsequently, string theories only work if you include higher dimensions.

In the original version of string theory, 25+1 dimensions were needed. The later string theories of the 1980s needed 9+1, and the most modern version needs 10+1. Once again we're going to need our Flatland analogy.

Imagine a being who lived on the surface of Flatland and observed a particle.

As far as they can tell, this particle is 2D, just as they themselves are. But if we looked at Flatland edge-on, we could see a third dimension, way too tiny for the Flatlander to notice.

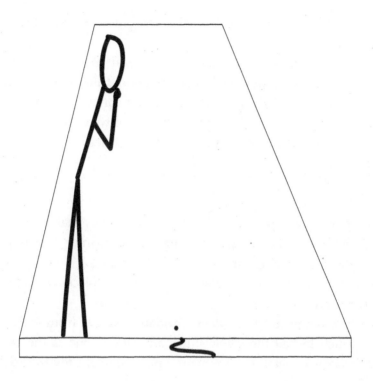

The particle observed by this Flatlander is actually the end of a string vibrating in a dimension too small to see. Our universe could very easily have a similar setup several dimensions higher. What we think of as particle could actually be vibrational modes of string stuff.

THE PROBLEM WITH STRINGS

At the moment string theory faces a minor setback. It doesn't work. Nobody actually has a string theory that describes all the known particles and their properties. What we have are a bunch of equations on the same theme, which describe a few aspects of our universe. None describes the whole thing.

There are so many different possible ways of assigning dimensions and vibrations to string theory that trying them all is practically impossible. A low estimate suggests there could be a minimum of 10^{500} possible versions of string theory (that's one with five hundred zeroes after it), only one of which would actually match our universe. Assuming any of them do.

Even if we set up a supercomputer to try a different version of string theory every second, it would take over a trillion, trillion, trillion years to try even a fraction of them. Finding the one correct version of string theory is like trying to find a specific droplet of water in all the Earth's oceans. During a storm. Blindfolded. With chopsticks.

THE STRENGTH OF STRINGS

You sometimes hear people saying that string theory makes no testable predictions so you can't do an experiment to confirm or deny it. (I hear that all the time in supermarket checkout queues.) A string theorist would counter by saying that there are three good pieces of evidence in the theory's favour, so far.

The first is the Regge trajectory evidence we get from early string theory, which kicked the whole thing off. The second is the fact that string theory does seem to have made a prediction already: the existence of gravity.

This is a more controversial claim because predicting the existence of something we already know about isn't a very good test of the theory. However, if Newton and Einstein had never formulated our theory of gravity and we only discovered quantum mechanics and particle physics, we might eventually have stumbled on string theory, which would predict gravity as one of its features. We would then carry out experiments to see if gravity was behaving the way the equations said and would discover, lo and behold, that it matches prediction.

The third success of string theory brings us back to the holographic principle and the information paradoxes we met earlier. Remember how Susskind and Maldacena said 3D information might be encoded two dimensionally on the event horizon of a black hole? Well, that wasn't quite what they found. I lied. Sorry. It's a bit more abstract than that.

First, let's revisit anti de Sitter space. That's the one where reality looks like a magnifying glass and things far away look compressed around the rim. Let's also stick to 2D for now.

Imagine a Flatland that has been curved into a horse-saddle shaped anti de Sitter (AdS) space. Now, imagine stacking another AdS Flatland on top of it. And then another, and another, until you have a column of AdS-curved Flatlands. I think the best analogy for this would be Pringles in a tube but I'm not sure of the legality of mentioning them. Unless, perhaps, I do a random plug for the product somewhere in the book.

Next, imagine these Pringles don't have gaps between them and can be merged to form a continuous tube of potato stuff. Most objects can't just be mashed together like that, but if we assume these Pringles are actually branes from string theory, they can.

In string theory, strings merge to form branes and branes can merge to form bulks, so if we take a bunch of Pringle-shaped string-theory branes (which should obviously be called Stringles) and stack them, we end up with one object rather than many layered ones. The outer surface of this object is a 2D cylinder and the inner volume is a 3D anti de Sitter space.

Now we just need to make two adjustments. First, these AdS Pringle membranes have to be infinitely big (this is theoretical physics, after all). That means the edge of each Pringle is an infinite distance away so we can never actually get to it.

Now, all you have to do is extend everything into a higher dimension again. In the calculation that makes the holographic principle work, the inner volume is actually a 4D AdS Pringle bulk and the outer 'surface' is a 3D Minkowski space.

We call a theory that works in Minkowski space a 'conformal field theory', or CFT for short. All of quantum mechanics needs to be a CFT but gravity and general relativity can happily work in AdS space.

This means our Pringle-tube has quantum mechanics taking place on the 3D CFT surface and gravity taking place inside the 4D AdS space. We call this linking between the two 'AdS/CFT correspondence' and this is how the holographic principle actually works.

Quantum information in a 3D CFT space can contain all the information about a 4D AdS space perpendicular to it, which is where gravity happens. So the holographic principle does not yet prove that 3D information can be stored two dimensionally. Instead, it says that 4D information can be stored three dimensionally.

Nobody has figured out a way to bring this theory down a dimension or make it work for regular space, as opposed to AdS space. This means, if we are honest, that the holographic principle has not quite solved the information paradox yet; it has just given us a clue that it might be possible to solve it using string theory one day.

Or Maybe It's Loops?

String theory is not the only game in town when it comes to uniting quantum mechanics with gravity or resolving the information paradox. Although it is a less well-known approach, the crushing difficulties of string theory have led many physicists to pursue an underdog instead: loop quantum gravity.

One of the many predictions of quantum mechanics is that energy (and therefore mass) comes in 'smallest possible units', which cannot be broken down. These units of mass are the fundamental particles of nature, sometimes called 'quanta'.

The problem with gravity comes when we start putting these fundamental particles onto the backdrop of spacetime. Particles need a non-curved environment for us to understand them, but spacetime is curved. So, asked John Archibald Wheeler, why not just chop space and time into smallest chunks as well?

Imagine there were such a thing as a 'smallest possible length', as if the Universe were pixellated on a computer screen rather than smoothed out like an oil painting. Any distance smaller than this minimum would be physically impossible, which is a conceptual headache. If a circle had a diameter of one of these minimum lengths, surely it ought to have a radius half of that?

Well, actually we have no reason to assume the laws of space at our scale behave the same at the quantum scales. We have already seen, several times, that just because an equation says something doesn't mean reality has to follow suit. Even an equation as basic as $1 \div 2 = \frac{1}{2}$ might not apply at every scale.

This hypothetical minimum length to our laws of physics can actually be calculated; it's called the Planck length (like the Planck mass) and has a value of 10^{-35} metres, written in longhand that's 0.00000000000000000 000000000000000001 metres. So how does this radical proposal of Wheeler's unite quantum mechanics with gravity? The answer is to do with how the Planck lengths are joined up.

If you look at a sheet of gauze netting, you can zoom right down to a scale where all the fibres are so tightly bound that they seem rigid. When we zoom back out the sheet can bend and fold, giving the appearance of flexibility, but really it is the angles between the fibres which change, giving the appearance of curvature.

In the same way, Planck-sized regions of space can be rigid and flat (as quantum mechanics requires) but the relationships between them are shiftable, meaning when we zoom out it can create the illusion of bendable spacetime (as general relativity requires). These tiny portions of quantum space form little loops just like the netting and thus the theory is called loop quantum gravity.

Loop quantum gravity has one major advantage over string theory in that it is much simpler mathematically. We do not have the complete thing worked out, but anything has to be better than searching for 1 among 10^{500} possible equations.

What's more, loop quantum gravity also makes a testable prediction (sort of). Because space has a smallest possible size, a shrinking black hole would eventually get to the point where the event horizon would have to shrink below this size, which is impossible. The only thing it could do would be expand back outwards, i.e. if loop quantum gravity is correct, black holes may not evaporate after all, nicely avoiding the information paradox.

Unfortunately, we would still have to wait trillions of years for the black holes near us to undergo this expansion process, during which time we may as well have run the string-theory calculation.

The other problem with loop quantum gravity is that when we incorporate time into the equations, which we inevitably have to do since bending space bends time simultaneously, it stops working.

This is actually where much of the disagreement between general relativity and quantum mechanics comes from – time doesn't work nicely in the quantum world. In order for loop quantum gravity to work, you therefore have to get rid of the concept of time altogether.

In loop quantum gravity there are different states of space existing but these relationships do not smoothly follow one from the other. In fact, there is a shortest unit of time allowed, the Planck time, lasting 10^{-43} seconds (0.0000000000000000000000000000000000000000001 seconds) and no event ever takes shorter or longer than that.

Time does not flow the way we imagine it to, rather it jumps from instant to instant creating the illusion of time the way individual frames played fast on a film strip create the illusion of motion. The details of how this happens are not at all clear and loop quantum gravity is thus woefully underdeveloped.

The physics of black holes is like Christmas come early for physicists. It's got everything we love: space, complexity, profound implications, counterintuitive conclusions and big fat dollops of weirdness on the side. Whatever theory we pick to account for them, we are forced to accept ideas that are truly alien. Which brings us nicely to the next part of the book.

PART III
LIFE AMONG THE STARS

CHAPTER EIGHT

Is Anybody Out There?

The Martyr

During the early Renaissance, Europe was dominated by Roman Catholicism and the separation of Church and State was non-existent. Debate was still encouraged but if a claim was at odds with the Vatican's interpretation of the Bible it could mean serious trouble, as we saw with Galileo.

One of the most extreme cases on record was that of the Italian monk and mathematician Giordano Bruno, who said in the sixteenth century that given the size of the Universe it was unlikely ours was the only inhabited world.

This suggestion could be seen as a challenge to certain passages of the Bible; most crucially, Romans 6:10 and Hebrews 7:27, which state that Jesus's death, which took place only once, was an atonement for all sin. If that is the case and aliens do exist then either:

1 aliens were not saved by Jesus's act, in which case God does not value every conscious being in creation;
2 aliens *are* saved, but are too far away to know about it, making Christianity unnecessary since salvation can be achieved without hearing the message of Jesus; or
3 the Bible made a mistake.

None of those explanations are particularly palatable if you're The Spanish Inquisition, so Bruno was tried for heresy. He was already a bit of a troublemaker who had a history of challenging Church doctrine, so it wasn't just his suggestion of alien life that got him in the clink, but regardless, he was found guilty of being a heretic and burned alive. Upside down. Naked.[1]

Today, there is a statue of Bruno on the site of his execution at Campo de' Fiori in Rome, marking him as a martyr for science (the statue is the right way up, by the way, oh and clothed). Fortunately, times have changed since the 1590s and I'm now allowed to speculate on the existence of alien life. I do quite like the idea of having my own statue, but I'd prefer not to leave this world naked and on fire.

GEEK FOR LIFE

The first movie I can remember watching as a child was the Leonard Nimoy-directed masterpiece *Star Trek IV: The Voyage Home* (1986), in which the crew of the Starship *Enterprise* have to travel back through time and rescue humpback whales in 1980s San Francisco, in order to help them communicate with an alien probe that only speaks whale. Seriously. It's one of the greatest movies ever made.

Star Trek remains one of my favourite franchises to this day and has always prided itself on depicting exotic forms of alien life. For example, in the episode 'The Tholian Web' (1968), the *Enterprise* crew encounter a species made entirely from crystal.[2] In 'Metamorphosis' (1967) they meet a sentient cloud of gas with Munchausen syndrome and in 'Sub Rosa' (1994) they meet an alien who takes the form of a Scottish sex ghost![3] Probably went a bit far with that one, in fairness.

The iconic line, 'It's life, Jim, but not as we know it,' is not from *Star Trek*, though. It's from the 1987 song 'Star Trekkin'' by the Firm, and the less said about that song the better. The closest actual quotation to this can be found in the episode 'Operation: Annihilate!' (1967), where Spock analyses a group of aliens that form a collective consciousness and says, 'It is not life as we know or understand it.'[4]

Oh, and if you're surprised by my obsessive attention to detail there you've obviously never met a Trekkie, and if you aren't comfortable with *Star Trek* references then I'm afraid you're in the wrong book.

TV shows such as *Star Trek* highlight the conceptual challenges we have when imagining alien entities. Life takes so many forms on our own planet that it's difficult to speculate what could be going on elsewhere. I mean, there's a species of caterpillar in South America called a velvet worm that fires hot glue from its face, and there are creatures called aeroplankton that live their entire lives drifting on air currents in the sky. Who knows what else is possible?

There could be creatures made of plasma living in the hearts of stars or civilisations made from spacetime nuggets living on the cusps of black holes. Perhaps we are being visited by aliens right now and just not recognising them for what they are.

Engaging in these speculations is always fun (and, I would argue, intellectually healthy) but asking, 'Could there be things we are unable to discover?' is answered simply by saying, 'Yes, and we will never know one way or the other,' which kills the discussion.

In order to address the question of alien life tangibly, we have to be a bit pessimistic. There absolutely could be life out there too strange for us to recognise, but in science we limit ourselves to what we can comment on. In other words, we need to assess the possibility of life as we *do* know it before we think about life as we don't.

WHAT IS LIFE?

Many workable definitions of life rely on compiling a list of characteristics that all living things share, e.g. consuming nutrition, responding to the environment, reproducing, moving, carrying out respiration, etc.

The problem with this approach is that we can always find exceptions. For instance, mules, hinnies, ligers, tigons, wholpins and pizzlies are hybrid creatures incapable of reproduction. Or take loriciferans, a species of microorganism found in the Mediterranean Sea that do not respirate. Clearly life means something a bit deeper than a list of common behaviours.

A friend of mine, in a philosophical mood, once suggested we define a living thing as simply 'that which can die'. Pretty clever, but sadly insufficient because there are some creatures that apparently don't. The plant species *Hydra viridissima* and the jellyfish species *Turritopsis dohrnii* do not age and, provided they have enough nutrition and are not actively murdered, seem to live forever.

The definition of life agreed upon by NASA at a 1994 conference on the subject was 'a self-sustaining chemical system capable of Darwinian evolution' and that's what most astrobiologists go by.[5]

The 'chemical system' bit narrows our definition to stuff made from atoms and molecules, ruling out life inside suns or black holes, etc. Next, 'Darwinian evolution' insists that these chemical structures have enough complexity to store and transfer information. What life therefore comes down to, essentially, is complicated chemistry and that allows us to make sensible guesses about what kind of planet or moon we should be looking for.

For starters, we need to find places that have the right ingredients, and there is good news here. Every organism on Earth is constructed from the same list of molecules – mostly amino acids assembled to form proteins (what cells are made of) and nucleotide bases assembled to form DNA (the instructions to make those proteins) – and these building blocks are not unique to Earth.

We have observed amino acids and nucleotide bases in far-off nebulae and found them locked inside crashed meteorites. The galaxy is apparently teeming with the building blocks needed for Earth-resembling life, and that's still being restrictive. Right here on Earth we have been able to create artificial DNA from unusual nucleotide bases (called XNA) and one scientist, Floyd Romesberg, has even created a semi-synthetic life-form with a new type of DNA altogether.[6]

Life can presumably form from a whole range of potential chemical starting points, so we just need to make sure a planet has the right conditions for reactions to occur.

Solid chemical systems such as rocks, metals and crystals are probably out because their atoms are fixed in rigid lattices and nothing moves, making interactions rare. By contrast, gases have the opposite problem because the particles are disordered and don't come into contact long enough for complex reactions.

For biochemistry to be feasible, the molecules need to be mobile, in close proximity and travelling at a low speed. It's therefore reasonable to suggest that life has the best chance of getting started if its chemicals are floating in a liquid. And not just any liquid.

It can't be too reactive or the compounds will react with the solvent rather than each other. It should also be capable of suspending many different types of molecule, and it should remain a liquid over a broad range of temperatures to accommodate seasonal changes on the planet.

There are only a few such chemicals known. Ammonia and hydrogen sulphide are good candidates but water is by far the most abundant. To be clear, there is no reason to insist that life *has* to have water, but water has all the right properties to give life a good shot.

On Earth the rule is that wherever we find water we find life inside it, and wherever we find life we find water inside it. Human beings, which we tend to think of as solid, are each made of around 37 trillion cells – each a tiny bag of watery stuff. We, like every living thing we know of, are walking, talking oceans and only certain planets can sustain something like that. A planet too close to its sun will evaporate its water systems, making life nearly impossible. Too far out and the opposite problem arises: everything freezes, halting the chemical reactions altogether.

For a planet to maintain liquid water it has to orbit its sun at a specific distance known as the circumstellar habitable zone, sometimes nick-named the 'Goldilocks zone' because its temperature, like the porridge, has to be just right.

How Many Goldilocks Planets Are There?

Let's consider our own galaxy, currently believed to contain 200 billion suns. Radiation near the galactic centre is pretty intense and would probably break down any complex biochemicals, so we'll be conservative and rule out the innermost quarter of the galaxy as inhospitable. That gives us 150 billion potential suns.

Next, we can assume most of these suns have planets, seeing as planet formation is a by-product of sun formation. Being pessimistic, though, we will assume only 90 per cent of suns have planets round them and only one planet per sun. That gives us 135 billion worlds.

The first exoplanets (planets orbiting a different sun to our own) were discovered in January 1992, given the names Draugr, Poltergeist and Phobetor, and since then we have logged over 3500 more.[7] I am reluctant to give a more precise number because we discover exoplanets at such a rapid rate that by the time this book goes to print, whatever number I write will probably be outdated.

Of the 3500 exoplanets we have found, 21 are suspected to be in Goldilocks zones.[8] That's 0.6 per cent of the total so, if we take this to be representative of the 135 billion planets in our galaxy, we get 810 million planets in Goldilocks zones. If we now assume only 10 per cent of these have water (and that's uncharitable seeing as we've already discovered a planet with liquid water called K2-18b[9]), that suggests 81 million planets where life could get a foothold.

This is where our extrapolations have to end, unfortunately, because we do not know how life originated on Earth, so we can't comment on whether it's a common or a rare process. But with 81 million habitable worlds (and that's pessimistic), even if life is a one in a million chance, there could still be up to 80 or so planets out there with life.

COULD THEY SEE US?

Suppose aliens were analysing our solar system and assessing it for life, just as we do with other solar systems. Would they be able to detect us?

The most important clue they would notice is the presence of oxygen in our atmosphere. Oxygen is a highly reactive element and doesn't last long in pure form because it usually gets incorporated into rocks and oceans. The only way a planet could maintain an oxygen atmosphere is if something on that planet is constantly producing it.

Furthermore, as a planet, Earth is not exactly discreet. We use a ton of radio communication, which leaks into space and anyone pointing a receiver at our corner of the galaxy could eavesdrop without difficulty. We have even sent a few messages into space on purpose, just in case someone out there is listening.

In 1974 we broadcast a three-minute signal towards the M13 star cluster from the Arecibo radio telescope in Puerto Rico, containing information about the chemical composition of our DNA. In 2001, the 'Teen Age Message', containing a theremin concert, was beamed towards six star systems (earliest to arrive in 2047, latest in 2070) and in 2008 the 'A Message from Earth' project transmitted 501 greetings from the social networking site Bebo at the planet Gliese 581c, due to arrive in 2029.

We've also included handy calling cards on the sides of our deep-space probes such as *Pioneer 10* and *11*, which left our solar system in 1983 and 1990 respectively. They were adorned with an aluminium plaque, coated in gold (one of the most stable metals), featuring a map of how to find Earth as well as pictures of a naked man and woman, in order to give aliens an idea what our more prudish civilians are offended by.

An even more elaborate message was devised for the two *Voyager* probes launched in 1977. Fixed to the hull of both those craft were golden phonographs containing 116 images of Earth, as well as recordings of common Earth sounds, a selection of music ranging from Beethoven to Chuck Berry, greetings in 55 languages and a sample of whale song. Y'know, just in case *Star Trek IV* was onto something.

What If Aliens Are Jerks?

The idea of reaching out to our cosmic brethren is an exciting one, but it has often been met with serious concern. If an alien species had hostile intent, sending them detailed information about us is basically briefing their invasion fleet on how to attack.

I mean, has nobody seen *Independence Day*? In that movie a race of aliens called the harvesters (there's your first freaking clue) destroys most of the human population within twenty-four hours, which is a plausible timeframe.[10]

A species would only have to be a few decades ahead of us in order to conquer us without any resistance. Imagine, for instance, if the armies of the 1970s declared war on the armies of the early 1900s. That would be a conflict of nuclear weapons, airplanes, radar, submarines, guided missiles and satellites against rifles and horses.

Obviously, in Hollywood movies, invading aliens always have an Achilles heel which we can exploit. In the movie *Signs* the aliens are allergic to water (logically invading a planet whose surface and atmosphere are full of it).[11] In *Mars Attacks!* it's the sound of yodelling that wins our victory, and in the aforementioned *Independence Day* the alien weakness turns out to be Jeff Goldblum.[12]

In reality, an alien race is unlikely to have a simple off switch, so maybe, some might say, we should be keeping our heads down before we become the next victims of a space-faring military. I understand where such a concern is coming from, but I would like to offer a counterargument, if I may.

All living things have to face the same basic challenges: compete for limited resources, ensure the survival of offspring, fend off rivals, etc. Aggressive tendencies help in achieving some of these goals but they are a hindrance for running a functioning society. An organism has to fight when threatened but it also has to withhold aggression towards family members and tribal allies. A species that does not develop the characteristics of restraint and patience will extinguish itself through in-fighting. There probably *aren't* any hyper-aggressive species out there.

Also, I think there's a good chance an alien species might even develop sympathy towards us. When the BBC broadcast its landmark series *Planet Earth II* in 2016, there was an outcry over one episode, which depicted baby turtles getting trapped in a drain.[13] Humans have no reason to feel pity for a different species and yet we do. We recognise something helpless and see it as pitiful rather than a threat.

My hunch is that empathy arises because a capacity for learning gives rise to a capacity for imitation and transposing our minds into others' points of view. A species that learns and thinks is likely to also develop some form of empathy for more pathetic lifeforms.

Also, from a pragmatic perspective, there simply wouldn't be much point in an alien invasion of Earth. A species capable of interstellar travel would be so technologically advanced that our measly planet would have nothing to offer. We have water, various minerals, a molten core and a sun, but those things are not in short supply – you can find them anywhere in the galaxy without a fight.

An alien species declaring war on us would be like us declaring war on penguins. They live far away from our comfortable home, they inhabit a place we don't want to live, it takes effort to get there, it has no resources we can use, they are not a threat to us, and we tend to find them cute. So there you have it, ladies and gentlemen. The human race. Penguins of the galaxy.

So, Where Is Everybody?

One afternoon in the summer of 1950, the great physicist Enrico Fermi was having lunch with his friends Edward Teller, Emil Konopinski and Herbert York, during which they made a similar calculation to the one we made a few pages ago. They concluded that even if life were a rare happenstance, the galaxy is big enough for it to still be very likely.

The conversation moved on, but Fermi went uncharacteristically quiet for a long time. Suddenly, he bellowed across the other three in mid-conversation, 'But where is everybody?'[14] It's sometimes called the Fermi paradox and presents astrobiology with a challenge: life should have arisen elsewhere yet we have never found it.

In the next chapter we will look at some of the intriguing clues that hint at possible alien life, but I'm afraid conclusive proof is not there. We are faced with the agonising question of why we seem to be alone even though the maths says we probably shouldn't be. There are many possible solutions to the Fermi paradox. Here are a few of the popular ones.

1 It could be that while alien life has emerged on other planets, it has remained aquatic. In order to develop technology you need to carry out chemistry because even the crudest metals have to be extracted from rocks, which requires mastery of fire. That tends to be hard to achieve underwater, so an intelligent sub-aquatic race will never alert us to their presence.

2 Even if alien life did make it to land, it might simply never develop technology. It only happened to one of the species on Earth so perhaps other worlds are inhabited by peaceful arboreal chimp creatures or feathery dinosaurs who never went extinct.

3 Perhaps there already are space-exploring civilisations, but the monumental distances between stars mean none of them have

reached us yet. Even travelling at close to the speed of light it would still take a hundred thousand years to cross the galaxy.

4 Maybe alien life has scouted us out already and considered us nothing worth paying attention to, ignoring us the way we would an ant colony.

5 Perhaps we come across as too aggressive and aliens want to see if we destroy ourselves via a self-imposed climate catastrophe before they risk a greeting.

6 Maybe alien civilisations are communicating with each other all the time but we haven't invented the technology they use. Right now text messages, phone calls and radio stations are moving through your body without you even noticing. In the same way, we could be sitting in the middle of a galactic social network so advanced the aliens would scoff at our primitive electromagnetic wave techniques. Don't we know about trans-quantum-mega-dynamo-fluctuals?

7 Perhaps the simplest explanation is that there is no paradox and we have overestimated the chances of life. Every species on Earth uses DNA, which implies that we share a common ances- tor, i.e. life only seems to have started on Earth once so it's not a very common event. Life could be a billion to one shot, in which case Earth really is the only inhabited place in the galaxy.

For my money, Arthur C. Clarke answered the Fermi paradox best in the following way: 'Two solutions exist, either we are alone in the Universe or we are not. Both are equally terrifying.'[15]

CHAPTER NINE

Greetings, Puny Earthlings!

SIGNALS IN THE GRASS

During the 1970s, in the south of England, intricate patterns started appearing in farmers' crops overnight. They were so precisely shaped and crafted, with interlocking right-angles and circles, that animals or weather phenomena had to be ruled out. This was the start of the crop-circle craze.

Roughly once every three weeks for ten years, a new pattern would appear somewhere in England and crop-circle experts (self-termed cereologists) began appearing as fast as the patterns themselves. Nobody knew who was creating the images, so naturally people turned to aliens as the chief culprit.

Until 9 September 1991 when the British newspaper *Today* ran a front-page story about two men named Doug Bower and Dave Chorley who admitted to being behind the crop circles. Chorley considered it a form of guerrilla art while Bower saw it as a simple prank, and rather than advanced alien-ray technology, they had been using planks of wood on the end of strings to create the designs. It was only when their project blossomed into mass hysteria that they were prompted to come forward and confess. The *Today* newspaper even brought in Pat Delgado, a noted cereologist, to view a recently formed crop circle, which he declared authentic, before the trickery was revealed. To Delgado's credit he accepted it like a true scientist and admitted that he had been fooled.[1]

The crop-circle craze is a reminder of how desperately everyone wants to believe in aliens and how easily we can fool ourselves into thinking we've discovered them. Extraterrestrial life would undoubtedly be the most exciting and significant discovery in history, but hopeful optimism can sometimes lead us to the point of accepting inadequate evidence. There is no shame in wanting there to be aliens, but we have to be on guard against wishful thinking.

THE TRUTH IS OUT THERE

In 1999 the independently funded French organisation *Comité d'Études Approfondies* (Committee for In-Depth Studies, COMETA) published a lengthy document with the aid of several members of the Institute of Advanced Studies for National Defence, called the COMETA report, which sought to assess UFO sightings over France. The report came to the conclusion that while most UFO sightings could be explained easily, some remained mysterious.[2]

It was far from an 'official report by the French government' as some have claimed (it was published as an article in the celebrity news magazine *VSD*, rather than a peer-reviewed journal) but it is an interesting read. We certainly can't explain everything we see in the sky so is alien visitation a reasonable explanation for these sightings?

At the Center for the Study of Extraterrestrial Intelligence (CSETI) Disclosure Conference in 2010, a number of US military personnel related their own stories about witnessing UFOs. One former US Air Force captain in particular, Robert Salas, even claimed that a UFO sighting he heard about coincided with the deactivation of nuclear guidance control systems at Malmstrom Air Force Base in March 1967 (although there is no evidence from anyone else at the base that it happened).[3]

We can go even higher up the chain, though. In 2014 the former Canadian Minister for Defence, Paul Hellyer, stated in an interview with *Russia Today* that he knew of a UFO incident that took place in 1961 in which objects were monitored flying over Europe, which he concluded to be alien spacecraft. He stated that aliens had been visiting Earth for thousands of years and were unimpressed with our stewardship of the planet.[4]

Hearing lunatics scream about flying saucers is one thing but hearing it from politicians and military personnel makes it seem more official somehow. Even former US President Jimmy Carter recounted seeing a UFO on one occasion so surely these kinds of testimonies carry some weight?[5] Well, not quite.

Politicians and military personnel witness strange stuff in the sky the same way civilians do. And, just like civilians, they are welcome to interpret them however they like. Their speculations are no more likely to be true just because their job puts them in a position of authority. Jimmy

Carter's UFO experience turned out to be the planet Venus, for instance, proving that even heads of state can be misled (and don't forget Jimmy Carter was a smart guy).[6]

I have occasionally run across conspiracists who point to testimonies such as these as evidence that there really are aliens visiting us, but to me that seems like putting faith in an authority figure, which is usually something conspiracists urge us *not* to do. Picking and choosing which government officials to trust based on whether they agree that we have been visited by aliens or not doesn't seem like a sensible way to conduct the discussion.

However, it is a matter of public record that many governments have kept details of UFO sightings. British intelligence chiefs founded the Flying Saucer Working Party in 1950 to investigate the matter and the Defence Intelligence Staff (DIS) carried out a three-year project between 1997 to 2000, called Project Condign, looking into UFO activity.[7]

Then, of course, there's the infamous Area 51 near the town of Rachel, Nevada – a highly secured military facility, which the CIA denied even existed for fifty-six years. The allure of this location is so great that in June 2019 2 million people signed a Facebook pledge that they would storm the base on 20 September under the logic that 'they can't stop all of us'.

It was initially started as a joke by gaming-streamer Matty Roberts who said that a big crowd 'Naruto-running' at the base (google it) would be faster than bullets, allowing everyone to 'see them aliens'. The 'Raid Area 51' meme eventually ballooned into a viral sensation so popular that the US Air Force had to issue a warning to anyone attending, explaining that guards were authorised to use whatever force was necessary to stop them.

Fortunately, less than 3000 people actually turned up in Rachel and only a third of that number made an approach to the gates, mostly to take selfies and chat with the guards. With the exception of a few brave souls who crossed the boundary line (and one brave idiot who urinated on it) nobody was arrested or injured.[8] But the very fact that the Area 51 conspiracy is so popular says something about how badly people want there to be proof of aliens.

What Are They Hiding?

We need to be healthily sceptical about claims of aliens and remember that a government keeping secrets is merely proof that they have secrets, not that those secrets have anything to do with extraterrestrial life. In fact, most politicians know the general public is crying out for confirmation of aliens and they know how much voter support they could earn by providing it. If 'secret alien evidence' really existed, the politician who came forward about it would never need to worry about election results again.

In fact, when a group of scientists thought they really had discovered aliens in 1996 (more on that shortly), President Bill Clinton publicly endorsed it and drew media attention to the claims.[9] Or consider the incident in 2017 when the Chilean government agency, CEFAA, deliberately released footage of a UFO to the public online to see if anyone could identify it.[10] Hardly an efficient cover up.

Besides, it's quite sensible for governments to keep files on UFOs. In recent years a lot of formerly classified documents on UFOs have been made public and the picture that emerges is one of espionage and security, not interstellar invasion. Governments want to know if they are being spied on by rival nations so investigating unfamiliar technology in the sky is worth doing.

In 2006, following a Freedom of Information Act request made by investigative journalist David Clarke, the DIS declassified their files on Project Condign, which showed that while UFOs were definitely a real thing (nobody doubts that) they had no more clue about what caused them than anyone else.

Even the clandestine Area 51 has now been unmasked, following a Freedom of Information request in 2005 by Jeffrey Richelson. It took a while for the request to be processed but in 2013 the CIA acknowledged the existence of Area 51 and were forced to declassify documents pertaining to what goes on there.[11]

There had been a previous attempt in 1994 to get the CIA to release information on Area 51, when a number of local residents claimed the military was using chemicals that affected their health. The government argued that all information relating to the base was classified and should not be released for the purposes of the court case because it would threaten national security. On that occasion Judge Philip Pro ruled in favour of the civilians' disclosure request but, ominously, President Bill Clinton intervened and issued a Presidential Determination, giving the base exempt status.[12] Richelson's request was more successful, however, and we now have a good idea of what's really going on there.

Firstly, it's no longer called Area 51 (and hasn't been called that since the 1970s). Its official designation is either Homey Airport or Edwards Air Force Base Detachment Three near Groom Lake (it may have other codenames but I don't want to dig too deep in case I get shot).

Operations at the base began in 1947 and what the military was doing was developing technology and spy aircraft, most prominently the OXCART and U-2 planes. It's worth noting that a U-2 plane is a very unusual shape and when it flies low during sunset the light reflecting off its wings gives it a disc-like appearance. Unsurprisingly, reports of UFOs in the Groom Lake area shot up at about the time U-2 test flights began.

It's also no surprise that the base had a lot of secrecy surrounding it. The military didn't want people to know anything about what they were developing and test pilots were given a brilliant cover story in case they crashed one of the planes. They were instructed to tell locals that their plane contained nuclear weapons and everyone should evacuate the area – a tactic that was used by pilot Ken Collins on 24 May 1963 after he crashed one of these $35 million planes. The area was evacuated due to nuclear bomb fear while a team from Area 51 swept in and removed details of the plane from the crash site – although bits of the wreckage can still be found there today.[13]

It's no wonder Area 51 got a reputation for secret clean-up jobs and weird vehicles in the sky. That really was going on. And the reason Bill Clinton had to silence the inquiry pertaining to Area 51's chemical output was because even knowledge of the chemicals being used at the base could tip off other countries about what kinds of tests were being conducted.

IT'S A BIRD . . . NO, IT'S A PLANE

As both the COMETA and DIS Condign reports found, UFOS are real but the thing which has to be remembered, and is all-too-often forgotten, is that the U stands for 'Unidentified', not 'Alien'.

The fact that governments have files on UFOs sounds dramatic because 'UFO' has become synonymous with 'extraterrestrial', but if we called them 'files on stuff in the sky we've got no clue about' it no longer sounds quite so sinister. Really, it's no surprise that aerial surveillance teams, both private and government funded, run across weird sky stuff all the time. The sky *is* weird.

The studies governments have carried out, such as Project Condign or Operation Blue Book (its American counterpart), have made careful analysis of UFO movements and show that they move at random rather than under any sort of guidance or piloting. Basically, governments are just as stumped about UFOs as everyone else. And that should not surprise us because UFOs are, by definition, things we cannot identify.

Claiming they are alien spacecraft is to identify them and in order to do that you need a reason, not just an unexplained observation. Simply saying 'I saw this thing in the sky that moved strangely' is not enough evidence to make the boldest claim in scientific history.

To show that extraterrestrial visitation is a good explanation for UFO sightings we would have to rule out more mundane explanations first and that's not easy. I mean, we don't even know how lightning works and we can barely predict the weather more than about twelve hours in advance. Saying we know enough about the sky to discount natural phenomena is not something we're even close to doing.

For instance, there is a recognised phenomenon called ball lightning, in which nuggets of glowing plasma go zipping through the air, chasing the electric fields that form inside clouds. Then there's the phenomenon of lenticular clouds, which is when clouds form saucer-like discs, often with glowing undersides illuminated by the sunset. Or take Fata Morgana, in which objects below the horizon appear to float above it due to atmospheric refraction.

We often don't even need to go that far to explain a UFO sighting. Take the Marfa lights, for instance. Between 1945 and 2008 over thirty sightings of mysterious lights moving across the horizon were observed near the town of Marfa, Texas. These lights were rightly classified as unidentified, until a study carried out by physics students in 2004 identified them as car headlights moving along Route 67 at an unusual angle from where the viewer stands.[14]

Or consider the Brown Mountain lights of North Carolina, which have been observed dozens of times since 1913 and were unidentified . . . until they was discovered to be the headlights of trains.[15]

Given that our senses are easily deceived and we barely understand how most of the stuff in the terrestrial sky works, it's premature to start ushering in aliens.

Besides, you would have thought that since the invention of smartphones the number of UFO sightings would have gone up dramatically. With everyone carrying portable video-recording devices in their pockets, if there really are lots of UFOs in the sky they should be recorded all the time. In actual fact, as camera quality has gone up, the number of UFO sightings has decreased . . . funny that.[16]

UFO peak season was the mid-1990s, a time when home video cameras were becoming popular but were still of poor quality. It's no mystery that dodgy recording equipment should correlate to objects in the sky that are hard to identify.

UFOs are real, but we can explain them one of two ways. Either they are unexplained phenomena we will eventually figure out, or they are alien craft going to the trouble of crossing trillions of kilometres at near the speed of light to visit our planet, just to hover in the sky for a few minutes. Seems like a wasted trip.

If aliens do want to contact us, flattening our crops and hovering in our atmosphere are not the best ways to go about it. If they want to make themselves known they would need to send us an obvious signal. Which, as it happens, they may have done.

Wow!

In 1855 something generated a radio transmission in the region of Tau Sagittarii, approximately 1 quadrillion kilometres away (0.6 quadrillion miles) from us. For over a century this signal crossed the cold vastness of space at the speed of light, before finally entering our atmosphere on 15 August 1977 where it was intercepted by the Big Ear telescope in Ohio.

For three days the transmission sat unnoticed on a data reel until astronomer Jerry Ehman printed off the week's readouts to see what the telescope had picked up. Sitting at his kitchen table on the 18th, Ehman perused the streams of numbers on dot-matrix paper and snapped upright with shock when he saw what was there.[17]

Radio signal intensity is assigned a number from 0 to 9, with 9 indicating the strongest signals. All stars give off a continuous flow of radio noise so most telescope data looks like a random string of 1s, 2s and 3s. What Ehman saw on his readout was as follows: 1111116EQUJ5111111. The signal was so intense it went beyond the 0–9 scale and had to use letters instead.

Seeing something like an A or a B would be intriguing for any astronomer, but to see a U-level transmission was unheard of. Ehman was so blown away that he circled the letters in red pen and wrote 'Wow!' beside them, giving it the name by which it is referred to today: the Wow! signal. And it wasn't just the intensity of the signal that caught Ehman's eye either. The specific radio frequency used is special.

Electromagnetic waves come in a practically infinite range, some as small as atomic nuclei and some as broad as planets. Given such a wide spectrum of wavelengths, which one should be used for sending signals between star systems?

You might think high-energy beams such as X-rays would be best but X-rays ricochet easily and rarely make it through a planetary atmosphere. The same is true of infrared and microwaves at the opposite end of the spectrum, being too low in energy to penetrate. Really, if you want to get a signal to the surface of a planet you want visible light or radio waves.

Visible light comes with the obvious catch of being easily blocked. You need a direct line of sight for it to work and anything as small as a moon or a dust cloud will absorb it. Radio waves, on the other hand, tend to be huge ripples in the electromagnetic field, often many kilometres long, so they tend to be diffuse and baggy, easily bending around objects or travelling through them unaffected.

That still leaves an enormous range of possible wavelengths to transmit on, so the most sensible thing to do would be to focus on wavelengths more likely to be recognised. We can assume, for instance, that an alien species smart enough to have mastered electromagnetic communication will know about pi, so the frequency of 3.14159 vibrations per second might be a good one to pick as that number would be significant to everyone, everywhere.

Another number we might reasonably assume aliens would be familiar with is 1420.4 million. That number may not seem important but it is immediately recognisable to anyone in astronomy. It is called the hydrogen line and represents the number of times a hydrogen atom will oscillate per second, i.e. hydrogen emits a radio pulse at a frequency of 1420.4 million hertz.

What's more, because all the stars in the galaxy are giving off a steady hum of 1420.4 million Hz, it's easy to filter out the background noise. If you get a broadcast on exactly that frequency with higher intensity than anything a sun would usually give off, you take notice.

The Wow! signal lasted for at least seventy-two seconds, which was the amount of time the Big Ear telescope was pointed at it. Annoyingly, it happened at night so nobody was awake to reorient the telescope so the signal could have gone on for hours and we will never know.

What we have in the Wow! signal is a transmission coming from another solar system bearing all the expected hallmarks of having been artificially generated and deliberately transmitted. And we only heard it once. If there was a natural explanation, then the signal would be constant or intermittent every time it gets generated by whatever is responsible. The fact we only got a single pulse is eerie.

However, we can argue that point in reverse. If the signal really was a deliberate attempt by some alien intelligence, why only the one? To date, the Wow! is the only signal of its kind and no matter how closely we have listened, the event has never been repeated.

Numerous physicists over the years have posed explanations for what might have caused the signal but none have stuck. We can assume, however, that if it really was a signal from a race of Sagittarians, it probably wasn't designed for us.

Any alien technologists studying our world 122 years ago when the signal was sent would only just be receiving light from Earth in 1733, when we had not even discovered the internal combustion engine. Aliens looking in our direction would see no indication there was a technologically advanced species due to emerge.

Perhaps the Wow! signal was a cosmic calling card – a beacon sent out at random in all directions to see if anyone answered. Perhaps the Sagittarians rotate their transmitters like a garden-sprinkler over the years and we are due for another pulse any day now. Perhaps it was just a bunch of Sagittarian grad students fooling around with the radio transmitter one night, making prank calls to the galaxy before their professor angrily kicked them out.

We have been listening for a repeat signal ever since and have heard nothing. Sagittarius remains silent.

We did send a reply in 2012, though. A collection of Twitter messages and a video by the comedian Stephen Colbert complaining about aliens anally probing abductees was encoded onto the hydrogen line and beamed towards Tau Sagittarii on 15 August, thirty-five years after we got the Wow!

If the Sagittarians are there and if they are listening, they will receive the message in 2134 and we will get a reply in 2256 – maybe this will be the first back and forth greeting between our world and another. Hopefully we'll get an answer about all the anal probes.

Martians Blew Up My Dog

On 27 December 1984 a group of researchers stumbled across a meteorite buried in the Allan Hills of Antarctica. The meteorite, designated ALH184001, was of Martian origin and dates to around 17 million years. It was most likely the result of a large meteorite impact on Mars spewing up surface rock, some of which made it into space before landing here.

Mars is in the Goldilocks zone of our sun and was believed to have oceans of liquid water 17 million years ago, right about the time the impact happened. Nothing was thought of this meteorite other than 'that's pretty cool' until a few years later when a research group led by David McKay started analysing rock fragments and found worm-shaped formations inside, looking similar to the bacterial fossils we have found on Earth.[18]

The story of the Allan Hills meteorite is reminiscent of the John Carpenter sci-fi horror *The Thing* in which an alien craft lands in Antarctica and terrorises a research outpost.[19] Fortunately, the 'Martian fossils' found in the rock were no bigger than a few micrometers and resemble segmented worms rather than shape-shifting dino-monsters. Still pretty cool, though. This was the aforementioned incident that compelled President Bill Clinton to make an international announcement that alien life had possibly been discovered.

The Allan Hills meteorite is not the only one to contain these fossil formations either. The Shergotty meteorite, which landed in India on 25 August 1865, contains structures that could also be indicative of bacterial life.[20] Then there's the Nakhla meteorite, which landed on a dog on 28 June 1911 in Egypt, vaporising it immediately (scientific honesty requires me to confess that there was only one eyewitness to this dog-splosion so the story may not be true). The Nakhla meteorite again has a few unusual fossil features that look similar to the kinds of bacterial traces we find on Earth.[21]

While some people suggested that the fossils found in the Allan Hills meteorite could have been caused by terrestrial bacteria (the meteorite landed on Earth 17 million years ago, after all), the Shergotty and Nakhla ones cannot be explained so easily – they were not around 17 million years ago when the ancient bacteria were doing their thing. Whatever caused the unusual worm-like shapes in these objects seems to have happened on the Martian surface and not the terrestrial one.

These fossils are not conclusive proof of alien life and many scientists have argued they could have formed via some unknown crystallisation process that occurred when the rock cooled from Martian magma. At the time of writing, the debate is not settled. They are shapes in the rock and shapes do not prove anything either way.

BUILD THE RING!

Our species gets most of its energy inefficiently. The Sun bathes our planet with an abundance of juicy light, which gets harvested by plants. These plants die, get buried and become compressed into carbon-dense minerals. We then take this wonderfully rich chemical feedstock and burn it.

Sometimes we use the burning directly to move engine components and sometimes we use it to heat boilers, the steam of which rotates magnets inside coils of copper wire, generating electric current. And that's what powers the whole show. The energy for all our vehicles, engines, machinery and electricity comes from burning dead plants and occasional bits of fish.

It's dirty. It's poisonous. It's disrupting the chemical balance of our atmosphere and even if you're not convinced by the overwhelming evidence for climate collapse, there is one indisputable fact about fossil fuel: it's going to run out eventually. This is a dumb way of doing things.

The Sun sits on our doorstep and kicks out enough energy to power the globe several times over, so if we can find a way to harness this energy directly, rather than relying on dead plants, we wouldn't have to worry about fossil fuels and all their associated problems. And there is a good chance another species will have come to that same conclusion.

If an alien species started the same way as us, using energy locked up in their plants, they will have faced the same crisis. An alien species that does not figure out a way of harnessing its sun's energy is probably not going to survive long enough to make contact. So they may hit on the same solution our sci-fi writers have done. Build megastructures in space to harness solar power.

The idea was first proposed by Freeman Dyson, who envisioned large rings and spherical arrays, now nicknamed 'Dyson structures', placed around the Sun. We might not be that far off building such structures ourselves, in fact. In October 2018, Wu Chunfeng, chairman of the Chengdu Aerospace Science and Technology Microelectronics System Research Institute in China, revealed his company's plans to build an artificial moon by the end of 2020.

Their idea is to create three enormous mirrors, which will go into orbit at an altitude of 500 kilometres (311 miles, a little higher than the

International Space Station) above the city of Chengdu in China. By reflecting sunlight towards the ground, the space mirror would be able to illuminate the city with a brightness eight times that of the Moon, saving 1.2 billion yen (approximately £132 million) a year in street lighting costs as well as cutting their fossil fuel usage.[22] Could extraterrestrials have similar ideas?

In 2015 we briefly thought we might have found a Dyson structure when astronomer Tabetha Boyajian released data on the star KIC 8462852 showing a regular pattern of dimming in the light as if there were large objects moving around it at regular intervals.[23] Despite the media coverage, Boyajian was highly sceptical she had really discovered a Dyson structure, and immediately sought to disprove her own hypothesis (she's one hell of a scientist). It turned out that the dimming effect was not purely opaque, which is what you'd get for a solid object, making the most likely candidate space dust.[24] Again.

To date, nobody has found a Dyson structure, nobody has picked up a clear transmission and nobody has discovered unambiguous alien fossils or corpses. If there is anyone out there, they are keeping quiet. Which means if we want to find out more about the Universe, we're going to have to go looking ourselves.

In the final chapter, therefore, I'm going to bring things back to us and investigate what we can do to take life to the Universe rather than the Universe bringing life to us.

CHAPTER TEN

Boldly Going

ACTUALLY, THIS *IS* ROCKET SCIENCE

The first rockets were invented in China, most likely by Tang Fu around the year 1000 CE for military and firework purposes.[1] Rockets work under the principle that powers most terrestrial vehicles – you take a chemical with a lot of energy packed into it and mix it with oxygen. Because oxygen is one of the most reactive elements on the periodic table, it will react violently, releasing the stored energy of the fuel in a process chemists call oxidation, more commonly known as an explosion.

Put this explosion inside a reinforced chamber with an opening that funnels the blast and Newton's third law does the rest, creating an equal and opposite force on the chamber to push it the other way. All you have to do is make sure it's pointed upwards.

It wasn't until 1944 that the German physicist Wernher von Braun built a rocket capable of reaching space: the V-2, from the German *Vergeltungswaffen*, which means 'retaliation weapons'. The Nazis fired over 3000 at city targets, and, in fact, von Braun's design was so efficient that modern rockets haven't deviated much.

The basic premise is not all that different to the internal combustion engine of a car. First, you need a chamber where you pack a bunch of fuel. Von Braun used alcohol and water while today's space agencies typically use kerosene or liquid hydrogen, both of which burn in the oxygen to generate water as a by-product.

Second, you need an oxygen tank and a mixing chamber where the fuel and oxygen are blended. Third comes the all-important combustion chamber with a nozzle at one end to direct the force of the blast. Ignite this fuel/oxygen combination and away you go.

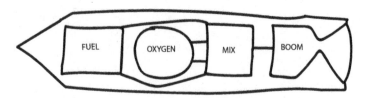

Essentially, a rocket is a carefully controlled explosion, which means things can go wrong if you make the slightest mistake, demonstrated perfectly on 14 April 1970 when the Apollo 13 mission's spacecraft had to abort its mission to the Moon because something exploded, 330,000 kilometres (205,000 miles) away from the safety of Earth.

The accident was the result of a series of mishaps you really couldn't make up. First, one of the oxygen tanks had previously been used for the Apollo 10 mission and during the process of transfer, it got dropped, falling a distance of no more than 5 centimetres (2 inches). From the outside everything looked fine, but that drop dislodged the tank's fill line – the tube running through the centre that feeds oxygen in and out.

During testing, an automatic heater inside the tank was supposed to boil out this excess oxygen but the heater switches had been built to run on 28 volts and at Kennedy Space Center they were using 65. This caused an overload of the switch circuits so the heater kept getting hotter and hotter without anyone realising. The heater was left on for a total of eight hours, during which the internal temperature rose to 540 °C, melting the Teflon coating of the wires that powered the tank's fan.

Fifty-six hours into the mission, the astronauts turned this fan on to churn up the liquid oxygen inside, ready for use. When they did so, the wires, which were now in direct contact with each other, short-circuited producing a spark . . . inside a tank that was leaking pure oxygen. Boom.[2]

Jack Swigert, piloting the craft, immediately radioed mission control in Houston and said, very calmly, 'I believe we've had a problem here.' Eight seconds later, Jack Lousma at mission control asked him to repeat the message and Jim Lovell, one of the other astronauts, responded with 'Uh, Houston, we've had a problem.' The third astronaut, Fred Haise, then elaborated: 'We had a pretty large bang,' to which Lovell added 'We are venting something out into space . . . it's a gas of some sort.'[3]

You have to admire the calm demeanour of the three astronauts in this situation. I'm pretty sure if I'd been on the mission the transcript would read something like, 'Oh my God! Holy %£$#@! The rocket's %£'@~#* exploded!' but then again, you don't get to be a NASA astronaut unless you've got the right stuff.

BRUTAL MATH

There are many reasons why rocket science has a reputation for being so complicated. The first is that everything in space is always moving, due to Newton's first law, which says objects move permanently in straight lines until an external force interferes.

On Earth, if you push something it will eventually come to a rest due to friction and air resistance but in space if you push on something it keeps moving forever. Nothing sits still outside our atmosphere, so your calculations have to take this into account as well as the fact that everything is going round in circles.

Isaac Newton proposed the idea of putting something into orbit by imagining a cannon so big it could fire a ball far enough to go round the world. The momentum of a cannonball tries to carry itself in a straight line, but gravity pulls it into a parabola instead.

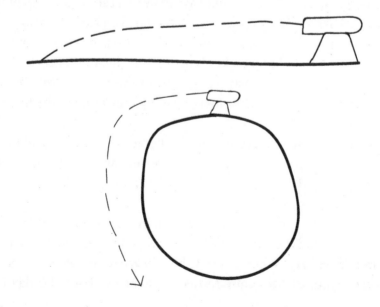

If we increase the momentum of the cannonball, however, we could fire it so far that by the time it fell, it would have overshot the planetary horizon, repeating the process again at the next horizon.

This is why astronauts experience weightlessness in space; it isn't because there's no gravity (there is), it's because their craft is trying to move in a straight line under momentum but the gravity of Earth is pulling them inwards. Astronauts are not really floating, they're falling. It's just that every time they fall towards the Earth, Earth has moved out of the way so they never land.

Also, rocket science is not just taking place in 2D like the diagrams above. It's happening in 3D space. Well, 3+1D spacetime really. And even that isn't the whole story because, surprisingly, most rocket scientists prefer to work in six spatial dimensions rather than three.

Every object in space needs to be described not just by location coordinates, but by momentum values as well, specifying how the rocket is moving in each direction. Rocket scientists act as if there are three 'momentum axes' a rocket is moving around in, i.e. when a rocket changes speed it is said to have changed its location in 'momentum space', a hypothetical grid of all the possible momenta it can take.

Then, of course, you've got to take into account the changing variables. To get a rocket into space you have to burn a certain amount of fuel but that fuel has a mass, meaning the more fuel you have the heavier the rocket becomes ... increasing your need for fuel. It would appear as though no rocket would ever take off because the heavier it gets the more fuel you need, making it heavier, etc. But in the plus column, as your rocket moves upwards it burns fuel, meaning it loses mass the higher it goes.

In rocket science everything is changing constantly, so although conceptually it might not be as difficult to picture as string theory or loop quantum gravity, the maths involved is a lot more complicated. And it's extremely easy to screw things up.

Never was this demonstrated better than in 1999 when NASA crashed their *Mars Climate Orbiter*, a $125 million probe sent to observe the Martian atmosphere. When a review board carried out an investigation into what happened, the culprit turned out to be two lines of code in the software that controlled the thrusters. The navigation team had been working in centimetres, while the contractors who built the probe had been using inches.[4]

Then again, who needs a space programme anyway? Some people argue that exploring the solar system is a waste of money when we have more pressing concerns on Earth. $125 million is a lot of cash just to find out what the air is like on Mars. I disagree vehemently with this viewpoint and I'll do my best to explain why in the next few sections.

COLLISION COURSE

In the 1998 Michael Bay sci-fi epic *Armageddon*, NASA learn of an asteroid headed for Earth which will eliminate all life, right down to the bacteria. Their solution? Send Bruce Willis and a team of oil drillers to land on the asteroid so they can nuke it from the inside out.[5] This is why sci-fi movies are better than other genres.

According to an internet legend that circulated after the movie was released, NASA began using *Armageddon* in its management training programme to see how many scientific inaccuracies their employees could spot. Sadly, this story turns out to be a myth, but what is true is that *Armageddon* raised public awareness of asteroids so much that there had to be an official government response to the outcry.

The US House of Representatives Science Subcommittee on Space and Aeronautics met on 21 May 1998 to address heightened public concerns over space research. They concluded that people wanted the government to take the threats of space seriously, which means Michael Bay's *Armageddon* may have done more to promote public awareness of science than any other film in history. Makes you wonder about *Transformers*, right?

The crucial thing to remember is that space is a dangerous place for a planet to be. Sixty-five million years ago, a piece of rock no more than 10 kilometres (6.2 miles) across hit the Earth, just south of what is now the Gulf of Mexico, creating an impact crater 180 kilometres (111 miles) wide and 20 kilometres (12.4 miles) deep, buried today beneath the Yucatan peninsula. When this rock struck, it not only cooked half the planet, it also sent up a plume of ash so big it blocked out our sunlight for years, killing most of the plants and subsequently the rest of the food chain.

Collisions like this happen all the time but our ever-moving tectonic plates and weather systems erode most of the evidence, giving us a false sense of immunity. A quick glance at the beautiful markings on the Moon, however, remind us that space is a shooting gallery. Those pretty patterns on the lunar surface are not decorations, they're battle scars.

Most of the time we are fortunate because big stuff rolling around the solar system is more likely to get pulled towards Jupiter. But if a big piece of rock happens to come barrelling towards us and Jupiter isn't in the right place, we will be in serious trouble.

Earth is struck by a meteorite, on average, every eighteen days. Most of these collisions are small, fortunately, and the largest near-impact in recent history is thought to be the Tunguska event (30 June 1908) when a lump of rock 190 metres (620 feet) wide made it through our atmosphere over eastern Siberia and destroyed 2000 square kilometres (772 square miles) of forest.[6] The figure sometimes given is that a large asteroid, like the one which killed the dinosaurs, should collide with us every 50 million years. It has been 65 million years since the last one, so are we overdue for another world-ending impact?

According to some astrophysicists, the risk of being hit by a large asteroid is lower today than it was in the past because the solar system has been cleared of its debris. All those markings on the surface of the Moon were made when the solar system was forming and things have since settled.

Other astrophysicists have argued that we have no way of making such a statement, because asteroids are small compared to planets, meaning they are harder to spot. Not to mention the fact that asteroids can be pulled out of their orbits and go careering off in unpredictable directions at any time.

Currently, there is only one institution responsible for monitoring near-Earth asteroids, called the Minor Planet Centre in Massachusetts. It provides daily updates on the locations of asteroids and plots the possible trajectories they could take.

Predicting collisions with accuracy is difficult because there are so many variables to factor in, and there are currently 900 objects on what is called the 'Sentry Risk Table' of potential impacts. The good news is that most of the asteroids we currently know of are low risk, but if you want to view the table yourself you can find it online at: https://cneos.jpl. nasa.gov/sentry/.

DESTROYER OF WORLDS

If you decide that the risk of asteroid collision is too minor to worry about, then fair enough. Personally, I think we should err on the side of caution when it comes to the end of the world, but I do take the point. It's probably worth my mentioning, however, that there are plenty of other ways in which space could destroy us.

Every few million years a powerful supernova, emitting high-energy radiation called a gamma-ray burst, occurs. One of those detonating within a few thousand light years would be enough to disintegrate our ozone layer, exposing us to cancer-causing ultraviolet light from the Sun. A gamma-ray burst is suspected to be the cause of the first mass extinction event 440 million years ago.[7]

Then there's the risk of rogue planets that have come loose from their own solar system and wander the galaxy untethered. If one of those entered our solar system it could gravitationally attract the other planets and pull them (including Earth) out of orbit.

Or, in the category of 'wouldn't kill us but would majorly suck', there is the threat of coronal mass ejections, where the Sun spits out plumes of charged particles from its surface which reach the Earth and wipe out our electrical equipment. The last time a major one hit us was in 1859, called the Carrington event after Richard Carrington, the astronomer who saw the explosion on the surface of the Sun moments before it reached us. Back then, it didn't cause too much disruption but the idea of losing GPS, the internet, bank records, the stock market and power networks all at the same time doesn't sound too good.

Then there are the increasingly pessimistic forecasts for our environmental stability. The human population is due to hit 10 billion by the end of the century, while at the same time climate collapse is making the habitable and arable regions scarcer. Simply put: more people, less space. Factor in the CO_2 already in the atmosphere, slowly dissolving into our oceans, acidifying them and killing the phytoplankton (which makes 50–85 per cent of our oxygen) and life on Earth is clearly becoming difficult.

And if that still isn't enough to concern you, play the really long game in which the Sun eventually expands and incinerates us. Any way you look at it, Earth is not eternal and neither is our guardian angel, Bruce Willis. Unless there are benevolent aliens out there who intend to swoop in and save us from extinction like in *The Day the Earth Stood Still*, we're going to have to handle these problems ourselves and that's why we need space programmes.[8]

We need to be developing technology and acquiring knowledge that will help us survive when (not if) our planet comes under threat. To quote Buzz Aldrin: 'We explore, or we expire.'[9]

The US government currently spends $21 billion a year on NASA research and while that might seem like a lot of money, remember the US public spends $210 billion a year on takeaway food.[10] Yes, we have enough things here on Earth that need financial attention, but the bigger picture is that if we aren't ready for challenges from space or challenges that threaten the planet as a whole, there won't be an Earth left.

That all sounds quite doom and gloom, I know, but fear not. Science is not just the tool by which we discover the problems, it's also the tool by which we develop the most promising solutions.

WHERE WE'RE GOING, WE DON'T NEED ROCKETS

Rockets are a good way of getting to orbit and reaching the inner planets, but even a fast one would be too slow to explore beyond the solar system. It took the *New Horizons* probe nine years to reach Pluto, travelling at 58,000 km/hour (36,000 mph). Going at that speed, it would take 78 millennia to reach the next solar system in Alpha Centauri. If we want to explore space properly we're going to need novel approaches to navigation.

One of the most efficient was an idea suggested in 1608 by none other than Johannes Kepler himself. He said that one day spaceships could be designed with solar sails fitted to catch the energy of sunlight rather than wind.[11] It's a darn good idea. So . . . we've built them.

On 26 June 2019 Elon Musk's company SpaceX launched LightSail 2, a satellite weighing only 5 kilograms, attached to an expanding solar sail with the surface area of a squash court, less than a human hair thick. By absorbing energy from the Sun this solar sail can be angled towards and away from the light, allowing it to move and change course without additional fuel.[12]

Even more exciting is the dream of Russian billionaire Yuri Milner, who has invested $100 million into the project Breakthrough Starshot, which plans to use solar-sail technology to reach Alpha Centauri in 20 years (as opposed to 78,000) by moving a craft towards it at 4 per cent the speed of light.

In order to do this, the craft will not be able to get enough energy from the Sun because it will be too far away, so Milner and his team are talking about building a 10 gigawatt array of lasers 800 metres (half a mile) wide in the Atacama Desert to propel the craft.[13]

And if that isn't cool enough, there's an even more outlandish idea being thrown around by physicists. Called the Alcubierre drive, it might be able to break the light-speed barrier altogether.

The principle was invented by the Mexican physicist Miguel Alcubierre who, as a young boy, would watch episodes of *Star Trek* in which the crew of the *Enterprise* travel faster than light using 'warp drive'. The writers of the show picked that term because it sounded snazzy but the concept of warping something to travel faster than light lodged in Alcubierre's mind and stayed there until he became a doctor of Physics at Cardiff University.[14]

Once Alcubierre learned about spacetime curvature in general relativity, he wondered if it might provide a loophole to the problem of light speed. The laws of physics forbid anything from travelling through spacetime faster than light, but what if the craft never moved at all?

Alcubierre proposed a scenario in which spacetime is contracted in front of the ship and expanded at the back, creating a gravity field at one end and an anti-gravity field at the other. By warping spacetime around the ship, you never have to worry about approaching light speed because you don't technically move at all. Space moves around you.

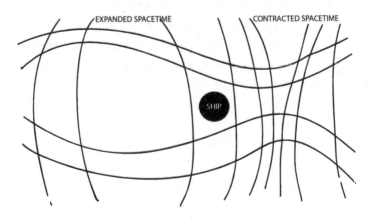

It sounds like sci-fi nonsense but NASA has reviewed Alcubierre's suggestion and found that the concept is sound. In a wonderful case of science fiction inspiring science fact, the Jet Propulsion Lab headed by Harold White concluded that if we could somehow generate a gravitational field and an inverse gravitational field around a ship, faster-than-light travel would really work.[15]

STAYING ALIVE IN SPACE

On 20 February 1947 the first living organisms were sent into space to see what the effects of cosmic radiation would be on biological tissue. In a captured V-2 rocket, American scientists from the White Sands Missile Range sent a box of fruit flies to a height of 109 kilometres (67.7 miles) before parachuting them back to the Earth.

The fruit flies seemed to be OK, with no sign of genetic abnormalities . . . so we started sending other creatures up to push the limits. In 1948 we sent up a rhesus monkey named Albert, who reached a height of 63 kilometres (39.1 miles) before sadly dying of suffocation. His successor, Albert II, made it all the way into space (134 kilometres/83.3 miles) but then died because the rocket's parachute failed to open.

Over the next few years the Americans and Soviets sent several monkeys, mice and a few dogs into space, including the famous Laika – the first dog to complete a full orbit aboard *Sputnik 2*, before dying due to heat.[16]

Then, on 12 April 1961, the USSR put the first human into space, Yuri Gagarin, an achievement that could only be topped by landing on the Moon itself, which the Americans achieved on 20 July 1969. So far only twelve people have walked the lunar surface, but many more have lived in Earth's orbit aboard the various space stations we have constructed.

The first, Salyut 1, was launched from the USSR in 1971 as a single unit, with the crew attempting to dock five days later. Unfortunately, the doorways failed to work and the crew had to fly back to Earth, which is the space-faring equivalent of driving to work, only to realise you left your work keys at home.

A month later the *Soyuz 11* craft had more success docking with Salyut 1 and its crew spent twenty-three days aboard, before they had to abandon ship due to a fire. Tragically, having escaped the fire on the station, the crew of *Soyuz 11* died on re-entry due to another technical malfunction.

Not to be deterred, however, the Soviets launched three more Salyut stations, operated by the military, and then the USA joined in with their own space station focused on scientific research, Skylab, in 1973. Again, this mission was beset with challenges because a tiny meteoroid damaged the lab during lift-off and the first crew had to spend their whole time getting power back online.

Then, in 1986, the USSR launched the MIR space station, which remained in orbit for fifteen years – so long that by the time the mission was complete, the USSR had dissolved and become modern-day Russia. MIR was also the first station to be constructed in orbit and the idea worked so well that the International Space Station (ISS) followed suit several years later.[17]

Sadly, the space race between the USA and the USSR slowly ground to a halt after all this excitement, with very little new development taking place for decades. This has been attributed to all sorts of things, ranging from money being needed elsewhere to a lack of public interest.

In fact, the surface of the Moon received no activity from 1976 to 2013 when the Chinese National Space Administration (CNSA) landed the Chang'e 3 mission's rover (named *Yutu*), the first soft-landing craft to reach the Moon in thirty-seven years. Since then, the CNSA has landed five more probes on the Moon including the Chang'e 4 in January 2019, the first ever to land on the far side.

The CNSA is fast becoming a major player in space exploration, having successfully put two prototype space stations in orbit – the Tiangong 1 and 2 – and plans for a full-scale station by the end of 2022. The CNSA has also stated that two of its long-term goals are to establish a permanent base on the Moon and eventually put people on Mars by 2040.

THE RED WORLD

As far as we are aware, Mars doesn't have life of its own but there is no reason we can't go there ourselves and, after decades of agonisingly slow progress, governments around the world have finally started to recognise the exciting opportunities a mission to Mars will bring. President Donald Trump even signed a document ordering NASA to land humans there by the end of 2033, seven years ahead of the CNSA.[18]

Entrepreneur Elon Musk also has a horse in the race, with SpaceX hoping to set up a Martian colony, beginning in 2022 with the launch of mining equipment that the colonists will use when they land. Jeff Bezos, the richest man in the world, has challenged Musk's confidence, however, saying that getting to Mars without establishing a Moon base is not going to be practical.

To that end, Bezos's company Blue Origin have unveiled plans to land on the Moon by 2024, setting up a base that we will use as a launchpad for future Martian missions. With superpowers and billionaires competing for interplanetary dominance, it looks like the space race may finally be getting underway again!

The challenges of getting to Mars are enormous, of course, both in terms of engineering skill and the biological effects it will have on the crew. Even if we pick a time when Earth and Mars are closely aligned, it could still take up to two years to make the trip, which would exact a hefty toll on the human body.

In order to test the effects of long-term isolation, the CNSA, the ESA (European Space Agency) and Roscosmos (the Russian space agency) ran an experiment in 2010 called Mars 500 in which six people were sealed in an artificial spacecraft for a year and a half to see how they coped. While no fighting or arguments broke out, the biggest challenge was that fatigue and psychological frustration set in after a few months as the crew adapted to their claustrophobic quarters.[19]

And once we arrive on Mars it isn't exactly going to be Eden. Mars has a gravitational pull roughly a third of that of the Earth and we have no idea what that will do to our bodies. Astronauts aboard the ISS have to undergo two and a half hours of exercise daily just to avoid muscle atrophy and bones overstretching.

It's also a cold planet, with maximum summer temperatures reaching –5 °C, not to mention the toxic air made of 95 per cent carbon dioxide. That atmosphere is also too shallow to pressurise our bodies and too thin to filter out ultraviolet rays from the Sun, meaning we would have to stay inside specially designed suits and habitats. On top of all that, Mars does not have a magnetic field to shield us from cosmic rays and solar winds so there might be a heightened risk of cancer.

Then again, we have already built bases for ourselves in Antarctica, on the hillsides of active volcanoes and, obviously, orbiting the Earth. We are an innovative species and we have a history of seeing obstacles as challenges so, if we want to go to Mars, we will.

People Like You and Me

During the Apollo programme, male astronauts had to urinate into a plastic bag adorned with a penis-receiver at one end. The size of this penis-receiver had to be correct, because if you picked one too small it would be too tight and if you picked one too large everything would leak . . . in a weightless environment.

The original penis-receivers were designated small, medium and large, but according to Michael Collins (who piloted Apollo 11) they were casually renamed 'extra-large', 'immense' and 'unbelievable' so that people would feel more comfortable picking the appropriate size.[20]

You might wonder why I have chosen such a puerile and silly anecdote to finish this book. It might seem a little anti-climactic but, to me, this story illustrates something we can easily forget. Astronauts have a reputation for being the finest specimens humanity has to offer but in reality they have insecurities, moments of self-doubt and senses of humour just like everyone else. Astronauts, like all scientists, are people.

Every single one of us alive today, no matter the circumstances of our upbringing or intellect, has moments of anxiety and unworthiness. Everyone wonders sometimes if the human race is worth preserving and everyone has moments where they look at the size of the Universe and feel insignificant by comparison. Space science can potentially be depressing because it reminds us of our smallness.

My answer is this: imagine if the Universe really *were* simple. Imagine how boring the story of science would be if Earth genuinely were flat, or if there was nothing outside our solar system. Imagine if, after spending a few years looking around with telescopes, we knew everything there was to know. No more mystery. No more exploration. No more discovery.

How awesome it is that instead we find ourselves in a universe as huge and varied as this one. How fortunate we are to be surrounded with so many mysteries in desperate need of solving and how lucky we are to live in a universe bigger than our imaginations.

All of us, every single one of us, is small, that's true. But all of us, every single one of us, has a thirst to understand where we fit in. We can all marvel at science and we can all take part in the adventure of learning what the cosmos has in store. Studying science is not reserved for elite boffins in lab coats with IQs in the 150s. Science is for everyone, with all our baggage, emotional hang-ups and self-doubts.

There is a whole reality out there waiting to be explored. And it will be. By people like you and me. Science is not simply the thing that brought us out of the caves: it is the thing that will take us to the stars. I truly believe, and always will believe, that science will save our species.

Appendices

Real Science Has Curves

Most traditional proofs for the roundness of the Earth rely on observing things such as the horizon or the stars, but Flat Earthers tend to be suspicious of these explanations. Here are three simple experiments you can carry out yourself, though, which show the Earth cannot be flat. I've discussed them with many a Flat Earther and have yet to hear a convincing counterargument.

1 If you heat a saucepan of water, energy from the flame is transferred to the water molecules, making them jump about. In order to fully evaporate the liquid, the molecules have to battle against the weight of air pressure above them, which means the more air you have overhead, the harder it is to get water to boil.

   If you do it at sea level, you will find that water boils at 100 °C. Any lower and the air pressure above will keep the water squashed into liquid form. But if you climb a mountain and try it again, you'll find the temperature drops roughly by one degree every 300 metres (328 yards).

   Logically this means you will eventually get to a point where water doesn't have to fight air above it and will boil with no effort at all – in other words the atmosphere must come to a stop.

   If the Earth were flat, the atmosphere would have to form a dome above it, highest at the North Pole in the centre and coming down around the edges at the South Pole. This would mean the further south you go the less air there is above you as the dome gets shallower. If Earth were flat, water should boil at lower temperatures in the south than in the north. Yet everyone from Denmark to South Africa finds that water in their saucepans boils at 100 °C.

The only explanation that matches both observations (boiling point remains constant with latitude but decreases with altitude) is that the atmosphere must be the same thickness everywhere, which can only happen if the Earth is spherical.

2 Gravity is a force that pulls inwards and crushes stuff into spheres, but many Flat Earthers deny gravity exists, believing that objects fall due to buoyancy differences with air or, more exotically, that Earth is accelerating upwards very fast, pushed from below as if on a giant elevator.

This is a creative way of explaining free-fall acceleration but it neglects a pretty easy experiment to carry out. Objects do not fall 'down' at all: they seek out the biggest mass in their vicinity. That's usually the Earth but not always. If you're standing next to something large, objects will actually fall a little bit sideways.

The simplest way to prove this is to stand beside a large mountain (the Schiehallion mountain in Scotland was the first to be used for this demonstration) and hang a pendulum beside it. What you will see is that the pendulum does not hang perfectly 'down' to the ground, it hangs ever so slightly askew, towards the mass of the mountain.

This demonstrates that gravity is all about masses attracting each other and not simply a 'downward pull'. Once you have established that gravity a) exists and b) pulls in all directions, it becomes inescapable that the Earth must be round since all its mass would self-gravitate into a clump. Even if it *had* been a disc at some point in the past, it would have been pulled together like a sheet of paper getting scrunched up.

3 This experiment will cost you a little bit of money, but the results are well worth it. First, get a digital camera and set it on a timer to take photographs every few minutes. Second, attach it to a high-altitude weather balloon (which cost between £20 and £500 depending on quality). Third, get permission from whoever

monitors air traffic in your country to release the balloon at whatever location and time they specify.

If you're good at techy stuff you can set the camera up to transmit the photographs to you on the ground. If you're not so good at the techy stuff, build a capsule for the camera with your contact details written on the side (and a small reward for anyone who finds it). As it rises higher and higher through the atmosphere, the camera will take a series of photographs, some of which will be high enough to capture the curvature of the Earth.

The reason I know this method works is because we carried it out at my school a few years ago. The camera not only recorded stunning photographs of its eight-hour journey, it managed to capture the Earth's curve at its peak. We currently have these photographs displayed in our Physics department corridor and I walk past them every morning.

Astrology for All

....

We call the repeating star constellations that pass overheard throughout a year the 'zodiac' and by measuring where we are in that cycle we can predict the rise and fall of the seasons. The Babylonians divided the zodiac into twelve slices, each named after whichever constellation happened to be behind the Sun (from our perspective) at that point. This is where we get the twelve 'star sign' names, e.g. Scorpio, Taurus, Gemini, etc.

Unfortunately, this base for astrology is pretty lousy. The Earth takes 365¼ days to circumnavigate the zodiac, which doesn't divide nicely into twelve. It divides far better into thirteen, so the sensible thing would have been to have thirteen months in the calendar. There's even a thirteenth major constellation called Ophiuchus, which would have fitted nicely, but the Babylonians really liked the number twelve so they ignored reality and fudged their numbers.

The constellations they chose were also pretty arbitrary and far from universally recognised. The ancient Chinese, for instance, split the zodiac into twenty-eight chunks, while the ancient Egyptians split it into thirty-six and the Mayans into nineteen.

The biggest problem facing astrology, however, is that the Earth gradually shifts with respect to the zodiac, meaning the twelve 'star signs' quoted in newspaper horoscopes are usually out of sync with what the constellations are actually doing.

As a result, roughly 86 per cent of people who claim they were born under one star sign were actually born under a different one altogether and have been reading the wrong horoscope for years.[1] Logically, reading the wrong horoscope should generate faulty predictions but this does not seem to be the case.

Most people who read their daily horoscope are very happy with it, for which I can think of only two explanations. Either the predictions are so generic and interchangeable that you cannot tell when you are reading the wrong one (meaning they are useless) or horoscopes are corresponding to some mystical phenomenon that splits the calendar into twelve uneven chunks. I will leave it to the reader to investigate.

The Grown-up Einstein Equation

Although he is probably best known for $E = mc^2$, the equation of Einstein's that gets talked about the most in astrophysics circles is this one:

$$R_{\mu v} - \frac{1}{2} R g_{\mu v} + \Lambda g_{\mu v} = \frac{8\pi G}{c^4} T_{\mu v}$$

Let's begin on the right-hand side of the equals sign.

- G: This is 'the gravitational constant', a number telling us how strong the force of gravity is in our universe. It has a value of $6.674 \times 10^{-11} \text{ m}^3\text{kg}^{-1}\text{s}^{-2}$.
- c: This is the universal speed limit; the speed at which gravity can move between two objects. It has a value of $3 \times 10^8 \text{ ms}^{-1}$.
- pi: This is 'pi', the number you get when you divide the circumference of any circle by its diameter. It has a value approximately equal to 3.14159 and shows up in loads of equations where forces (such as gravity) extend outwards spherically in all directions.
- T: This is called the 'stress-energy tensor'. A tensor is a number grid that has values at each point inside it telling us what is going on at each coordinate in a given space. Tensors are useful for describing what's happening in 3D, so although it looks like a single number, T actually represents lots of them. Specifically, the stress-energy tensor tells us how much energy (or mass) there is at every given point in the portion of spacetime we are investigating.
- μv: The little μv things throughout the equation are just a notation used for defining the coordinates in a given space and they tell

us how things change when we move from one point to another. You see them a lot in the physics of fields because we're talking about how our surroundings change as we move around in 3D.

Now let's move to the left-hand side of the equation.

R: There are two R symbols and annoyingly they are not representing the same thing (thanks, physicists). The $R_{\mu\nu}$ symbol at the start is called 'the Ricci curvature tensor'. The 'Ricci' bit refers to the Italian mathematician Gregorio Ricci-Curbastro, the 'curvature' bit is referring to how much the geometry of spacetime is curving, and the 'tensor' bit is once again a number grid telling us about all the changing points and directions inside the spacetime region. The second R (in front of the g) is a simpler term. It's not a tensor this time, it's just a single number that tells us how curved the spacetime is. An analogy might be to picture a room in which temperature is fluctuating wildly everywhere inside it. We might pick a point in the room to study and look at a tensor that tells us how the values are changing up and down at every point, but we could also take a simpler measurement of 'what single number value is the temperature right now?' That's the difference between the two Rs.

Λ: This is called 'the cosmological constant' and it describes antigravity. Because Einstein's equation initially seemed to predict that everything should attract, he added this term to explain why the Universe is not collapsing. He later removed it when he found out the Universe is expanding due to the big bang and now a lot of physicists have put it back in, having discovered dark energy (see Chapter Five).

g: This is called the 'metric tensor' and is perhaps the heart of the equation. This tensor is a number grid that defines how bendable spacetime actually is. While the R terms and the T terms depend on the scenario we are studying, the g term is a constant and links the two together. What it tells us is how much a certain amount of energy will warp a certain region of spacetime.

If we go back to our room-temperature analogy, the T term is like describing a fireplace or a hot stove, i.e. the source of the temperature distortion; the R terms are describing what the air temperature is actually doing; and the g term tells us how heat and air temperature are always linked. We can take this g term and apply it to any sized room (any R terms) and introduce any kind of heat source (any T terms) and g will tell us how they interact.

All in all, this equation is telling us that if we take the amount of energy in a given region of space, factor in gravity and the speed at which gravity can move (right-hand side of the equation), we can calculate how the dimensions of space and time will be warped as a result (left-hand side of the equation).

The Mysterious Origins
of Hawking Radiation

There are quite a few explanations for how Hawking radiation might occur. I'm going to go with the one Hawking himself proposed, because it seems only fitting. It's also the only one I vaguely understand.

In Chapter Five we saw that particle pairs can pop into existence from a beam of light before vanishing an instant later. The full process has many steps to it (if you're curious, I devote a whole chapter to it in my book, *Fundamental*) but the basic premise is simple. Particle pairs can appear in empty space for a fraction of a second, before disappearing back into it.

Most of the time this doesn't cause a problem for the Universe because the amount of energy stays constant overall. A particle pair can exist, created out of nothing, but then it has the good manners to vanish just as quickly, so the Universe gets its energy back.

But on the event horizon of a black hole things might be different. If a particle pair pops into existence on the surface boundary, one of the particles would fall into the black hole but the other would make it out. There is now a new particle flying away from the black hole, apparently containing energy that wasn't there before. This is the Hawking radiation.

The only way to avoid creating energy from nothing would be for the in-falling particle to acquire a kind of 'negative energy' as it hurtles towards the interior, cancelling out whatever's in there. Thus, part of the black hole is eroded away and the total energy of the Universe is restored. It's as if the Universe has some kind of accounting system where the balance always has to be the same at the end of the day. In fact, let's run with that as an analogy.

The Universe is a bit like a bank, inside which there sits a disgruntled employee called Little Miss Black Hole. On her desk is a pile of particles she has been accumulating over the years and is very proud of. It doesn't belong to the bank of the Universe, she just likes to keep it within arm's reach.

One day, in comes Jolly Mr Space, who borrows a particle of his own from the bank. The bank now has a balance of −1 particles, while Jolly Mr Space has +1. Under normal circumstances Jolly Mr Space will return his particle to the bank a moment later and the balance resets, but things get complicated when Little Miss Black Hole is involved.

In a fit of rage, when Jolly Mr Space borrows his particle from the bank, Little Miss Black Hole snatches it from him and hurls it out of the window. Jolly Mr Space is now in a bit of a pickle. The bank is down −1 particle and demands he pay it back, but he doesn't have the particle he borrowed any more, it's flown off somewhere else.

At this point, however, Little Miss Black Hole has an attack of conscience and realises that causing this problem wasn't fair. In order to get space out of trouble, she grudgingly offers to hand over one of her own private particles to the bank. That way, the debt is restored. The bank's balance returns to zero and Jolly Mr Space doesn't have to worry about the particle that went flying out the window. The only one who has lost out here is Little Miss Black Hole, who had to lose some of her own matter in the process.

The bank doesn't really care that it got its particle back from Little Miss Black Hole rather than Jolly Mr Space, it just cares that the balance has been restored. And if this happens for a very long time, then eventually Little Miss Black Hole will run out of particles and vanish forever.

Acknowledgements

Astronomical represents, for me, a five-year writing journey and, following Elemental and Fundamental, the completion of a science-book trilogy. I get asked a lot what got me into writing and the more you get asked a question like that, the more you're forced to think about it. Ultimately, I think the point of writing is pretty straightforward: give something positive to the people around you. I've been extremely lucky to have the following people around me . . .

First, I want to thank BreeAnne Kelly, to whom the book is also dedicated. BreeAnne's favourite topic is space and I wrote Astronomical while trying to picture her reaction to it. I hope it lives up to her endless enthusiasm.

Second, I want to thank Revd. Lloyd Southgate, not only for his encouragement and feedback on the text of the book, but for giving me so much inspiration as a science educator. Peace and blessings be with you.

Third, I want to thank Karl Dixon, who regular readers may notice appears in the acknowledgements of all three books. Karl has been one of my biggest supporters since day one of writing (and I don't just mean professionally). I owe him an enormous creative debt.

Fourth, I want to thank my agent Jen Christie for always challenging my ridiculous ideas and bringing me down to Earth so that I write something actual humans want to read.

Fifth, I want to thank Duncan Proudfoot, Amanda Keats, Howard Watson, Beth Wright and everyone else at Little, Brown, with whom it has been such a delight to work. I had no idea people in the publishing industry were such a cheerful, engaging and dedicated bunch.

Sixth, I want to thank Andrew Miles for checking my physics and correcting me when I was getting it wrong. If the science in the book is accurate, thank him. If it's wrong, blame me.

Seventh, I want to thank anyone and everyone who has ever worked on a *Star Trek* film, TV series, novel, comic book, magazine, musical, leaflet, postcard, sandwich, etc. I was reading/listening to/watching *Star Trek* media during the whole process of writing and it kept my spirits up during every sentence.

And lastly, I want to thank my father, as well as all the other parents out there who go star-gazing with their children and show them the magic of the Universe.

References

Introduction

1. Lauren Said-Moorhouse, 'Rapper B.o.B. thinks the Earth is flat, has photographs to prove it', CNN (26 January 2016). Available at: https://edition.cnn.com/2016/01/26/entertainment/rapper-bob-earth-flat-theory/ (accessed 13 September 2019).

2. Alex Knapp, 'The lyrics to B.o.B.'s flat earth anthem "Flatline" with science annotations', *Forbes* (26 January 2016). Available at: https://www.forbes.com/sites/alexknapp/2016/01/26/the-lyrics-to-b-o-b-s-flat-earth-anthem-flatline-with-science-annotations/#4e792aa455d4 (accessed 13 September 2019).

3. C. Garwood, *Flat Earth: The History of an Infamous Idea* (London: Pan, 2008); A. R. Wallace, 'The rotundity of the Earth', *Nature*, vol. 1, no. 23 (1870), p. 581.

4. U. G. Morrow, *The Earth a Concave Sphere* (Estero, FL: Guiding Star, 1905).

5. Hoang Nguyan, 'Most flat earthers consider themselves very religious', YouGov (2 April 2018). Available at: https://today.yougov.com/topics/philosophy/articles-reports/2018/04/02/most-flat-earthers-consider-themselves-religious (accessed 13 September 2019).

Chapter One

1. C. Sagan, *The Dragons of Eden* (New York: Random House, 1977).

2. P. A. Oesch et al., 'A remarkably luminous galaxy at Z=11.1 measured with the Hubble Telescope GRISM Spectroscopy', *Astrophysical Journal*, vol. 819, no. 2 (2016) pp. 465–73.

3. B. Sato et al., 'The N2K Consortium. II.A Transiting hot Saturn around HD 149026 with a large dense core', *Astrophysical Journal*, vol. 633, no. 1

(2005), pp. 465–73; A. Muller et al., 'Orbital and atmospheric characterization of the planet within the gap of the PDS 70 transition disk?'https://iopscience.iop.org/article/10.3847/0004-637X/819/2/129, *Astronomy and Astrophysics*, vol. 617 (September 2018).

4. S. Kaplan, 'Why a tiny Lego version of Galileo rode on NASA's Juno probe all the way to Jupiter', *Washington Post* (5 July 2016).

5. N. Madhusudhan et al., 'A Possible carbon-rich interior in super-earth 55 Cancri e', *Astrophysical Journal Letters*, vol. 756, no. 2 (2012).

6. M. A. Kenworthy and E. E. Mamjek, 'Modeling giant extrasolar ring systems in eclipse and the case of J1407b: sculpting by exomoons?', vol. 800 (2015).

7. 'Hubble finds a star eating a planet', NASA (20 May 2010). Available at: https://www.nasa.gov/mission_pages/hubble/science/planet-eater.html (accessed 13 September 2019).

8. A. Legar et al., 'The extreme physical properties of the CoRoT-7b super-Earth', *Icarus*, vol. 213, no. 1 (2011), pp. 1–11; L. A. Rogers and S. Seager, 'Three possible origins for the gas layer on GJ 1214b', *Astrophysical Journal*, vol. 716, no. 2 (2010), pp. 1208–16.

9. Richa Gupta, 'Raspberries and rum – Sagittarius B2', *Astronaut* (12 August 2015). Available at: https://astronaut.com/raspberries-and-rum-sagittarius-b2/ (accessed 13 September 2019).

10. R. Sahai, S. Scibelli and M. R. Morris, 'High-speed bullet ejections during the AGB-to-planetary nebula transition: HSI observations of the carbon star, V Hydrae', *Astrophysical Journal*, vol. 827, no. 2 (2016), p. 92.

11. D. M. Kipping and D. S. Spiegel, 'Detection of visible light from the darkest world', *Monthly Notices of the Royal Astronomical Society: Letters*, vol. 417, no. 1 (2011), pp. 1–5.

12. M. Konacki et al., 'An extrasolar planet that transits the disc of its parent star', *Nature*, vol. 421 (30 January 2003), pp. 507–9.

13. D. J. Armstrong et al., 'Variability in the atmosphere of the hot giant planet HAT-P-7b', *Nature Astronomy*, vol. 1, no. 1 (2016).

14. 'Rains of terror on exoplanet HD-189733b', NASA (31 October 2016). Available at: https://www.nasa.gov/image-feature/rains-of-terror-on-exoplanet-hd-189733b (accessed 13 September 2019).

Chapter Two

1. M. J. L. Young, *Religion, Learning and Science in the Abbasid Period* (Cambridge: Cambridge University Press, 2006).

2. J. Needham and C. Ronan, 'Chinese cosmology', in Norriss S. Hetherington (ed.), *Cosmology: History, Literacy, Philosophical, Religious and Scientific Perspectives* (New York: Garland, 1993).

3. N. Oresme, *The Book of the Heavens and the Earth*, trans. A. D. Menut and A. J. Denomy (Madison, WI: University of Wisconsin Press, 1968).

4. S. Weinberg, *To Explain the World: The Discovery of Modern Science* (London: Allen Lane, 2015).

5. D. H. Kobe, 'Copernicus and Martin Luther: An encounter between science and religion', *American Journal of Physics*, vol. 66, no. 3 (1998), p. 190.

6. W. J. Boerst, *Tycho Brahe: Mapping the Heavens* (Greensboro, NC: Morgan Reynolds, 2003); H. Håkansson, *Letting the Soul Fly among the Turrets of the Sky* (Stockholm: Atlantis, 2006); A. Wilkins, 'The crazy life and crazier death of Tycho Brahe, history's strangest astronomer', *io9* (22 November 2010). Available at: https://io9.gizmodo.com/the-crazy-life-and-crazier-death-of-tycho-brahe-histor-5696469 (accessed 14 September 2019).

7. J. Kepler, *Epitome of Copernican Astronomy and Harmonies of the World*, trans. C. G. Wallis (Amhurst, NY: Prometheus, 1995).

8. Queen, 'Bohemian Rhapsody', written by Freddie Mercury (EMI, 1975).

9. Robert Bellarmine, letter to Paolo Foscarini, entitled 'Letter on Galileo's Theories' (12 April 1615).

10. S. Drake, *Discoveries and Opinions of Galileo* (New York: Anchor, 1957).

11. A. Dreger, *Galileo's Middle Finger* (London: Penguin, 2017).

12. 'Vatican admits Galileo was right', *New Scientist* (7 November 1992).

13. C. Peebles, *Asteroids: A History* (Washington, DC: Smithsonian, 2001).

14. P. Rincon, 'The girl who named a planet', *BBC News* (13 January 2006). Available at: http://news.bbc.co.uk/1/hi/sci/tech/4596246.stm (accessed 4 October 2019).

15. D. Jewitt and J. Luu, 'Discovery of the candidate Kuiper belt object 1992 QB_1', *Nature*, vol. 362 (22 April 1993), pp. 730–2.

16. C. Trujillo and S. S. Sheppard, 'A Sedna-like body with a perihelion of 80 astronomical units', *Nature*, vol. 507 (27 March 2014), pp. 471–4.

17. B. Guarino, 'New dwarf planet spotted at the very fringe of our solar system', *Washington Post* (2 October 2018).

18. J. Scholtz and J. Unwin, 'What if Planet 9 is a primordial black hole?', unpublished (24 September 2019). Available at: https://arxiv.org/abs/1909.11090 (accessed 4 October 2019).

CHAPTER THREE

1. US Patent 1-781-541, issued 11 November 1930; R. Greenfield, 'Celebrity invention: Albert Einstein's fancy blouse', *Atlantic* (22 April 2011). Available at: https://www.theatlantic.com/technology/archive/2011/04/celebrity-invention-albert-einsteins-fancy-blouse/237704/ (accessed 14 September 2019).

2. J. P. Luminet, *Black Holes*, trans. A. Bullough and A. King (Cambridge: Cambridge University Press, 1987).

3. J. Wheeler, K. Thorne and C. W. Misner, *Gravitation* (Princeton, NJ: Princeton University Press, 1987).

4. A. Aczel, *God's Equation: Einstein, Relativity and the Expanding Universe* (New York: Delta, 2000).

5. G. Lemaître, 'The beginning of the world from the point of view of quantum theory', *Nature*, vol. 127 (9 May 1931), p. 706.

6. S. Singh, *Big Bang* (London: HarperCollins, 2010).

7. S. Mitton, *Fred Hoyle: A Life in Science* (Cambridge: Cambridge University Press, 2011).

8. R. A. Alpher and R. C. Herman, 'On the relative abundance of the elements', *Physical Review*, vol. 74, no. 12 (1948), pp. 1737–42.

9. A. G. Levine, 'The large horn antenna and the discovery of cosmic microwave background', American Physical Society (2009). Available at: https://www.aps.org/programs/outreach/history/historicsites/penziaswilson.cfm (accessed 14 September 2019).

Chapter Four

1. S. Hall, 'BICEP2 was wrong, but sharing the results was right', *Discover Magazine* (30 January 2015).

Chapter Five

1. F. Nicastro et al., 'Observations of the missing baryons in the warm-hot intergalactic medium', *Nature*, vol. 558 (21 June 2018), pp. 406–9.
2. Eminem, 'Without Me', written by Marshall Mathers, Jeffrey Bass, Kevin Bell, Anne Dudley, Malcolm McLaren and Trevor Horn, Interscope Records (2002).
3. G. Gamow, *My World Line: An Informal Autobiography* (London: Viking Press, 1970).
4. A. G. Riess et al., 'Observational evidence from supernovae for an accelerating universe and a cosmological constant', *Astronomical Journal*, vol. 116, no. 3 (1998), pp.1009–38; S. Perlumutter et al., 'Measurements of the Omega and Lambda from 42 high-redshift supernovae', *Astrophysical Journal*, vol. 517, no. 2 (1999), pp. 565–6.

Chapter Six

1. D. V. Martynov et al., 'Sensitivity of the advanced LIGO detectors at the beginning of gravitational wave astronomy', *Physical Review D*, vol. 93, no. 11 (2016).
2. S. Schaffer, 'John Mitchell and black holes', *Journal for the History of Astronomy*, vol. 10 (1979), pp. 42–3.
3. M. Bailes et al., 'Transformation of a star into a planet in a millisecond pulsar binary', *Science*, vol. 333, no. 6050 (2011), pp. 1717–20.
4. C. M. Zhang et al., 'Does submillisecond pulsar XTE J1739-285 contain a weak magnetic neutron star or quark star?', *Publications of the Astronomical Society of the Pacific*, vol. 119, no. 860 (2007), p. 1108.
5. S. Doeleman, 'EHT status update, December 15 2017', Event Horizon Telescope (15 December 2017). Available at: https://eventhorizon-telescope.org/blog/eht-status-update-december-15-2017 (accessed 14 September 2019).

6. *Interstellar*, written by Jonathan Nolan and Christopher Nolan, directed by Christopher Nolan, Paramount Pictures/Warner Bros. Pictures (2014).
7. K. Thorne, *The Science of Interstellar* (New York: W. W. Norton, 2014).
8. O. James et al., 'Gravitational lensing by spinning black holes in astrophysics, and in the movie *Interstellar*', *Classical and Quantum Gravity*, vol. 32 (2015).
9. C. W. Misner and J. A. Wheeler, 'Classical physics as geometry', *Annals of Physics*, vol. 2, no. 6 (1957), p. 525.

Chapter Seven

1. O. Lahav et al., 'Realization of a sonic black hole analog in a Bose–Einstein condensate', *Physical Review Letters*, vol. 105, no. 24 (2010).
2. Alan Lightman, 'The day Feynman worked out black hole radiation on my blackboard', *Nautilus* (11 April 2019).
3. S. Hawking, 'Into a black hole', Hawking.org. Available at: http://www.hawking.org.uk/into-a-black-hole.html (accessed 14 September 2019).
4. T. Yoneya, 'Connection of dual models to electrodynamics and gravidynamics', *Progress of Theoretical Physics*, vol. 51, no. 6 (1974), pp. 1907–20; P. C. W. Davies and J. Brown, *Superstrings: A Theory of Everything?* (Cambridge: Cambridge University Press, 1988).

Chapter Eight

1. T. Fitzgerald, *Discourse on Civility and Barbarity* (Oxford: Oxford University Press, 2007).
2. 'The Tholian Web', *Star Trek*, season 3, episode 9, written by J. Burns, C. Richards, directed by H. Wallerstein, NBC (15 November 1968).
3. 'Metamorphosis', *Star Trek*, season 2, episode 9, written by G. L. Coon, directed by R. Senensky, NBC (10 November 1967); 'Sub Rosa', *Star Trek: The Next Generation*, season 7, episode 14, written

by B. Braga, directed by J. Frakes, Broadcast Syndication (31 January 1994).

4. The Firm, 'Star Trekkin'', written by John O'Conner, Grahame Lister and Rory Kehoe, Bark Records (1987); 'Operation: Annihilate!', *Star Trek*, season 1, episode 29, written by S. W. Carabatsos, directed by H. Daugherty, NBC (13 April 1967).

5. 'About life detection', NASA. Available at: https://astrobiology.nasa.gov/research/life-detection/about/ (accessed 14 September 2019).

6. D. A. Malyshev et al., 'A semi-synthetic organism with an expanded genetic alphabet', *Nature*, vol. 509 (15 May 2014), pp. 385–8.

7. A. Wolszczan and D. A. Frail, 'A planetary system around the milli-second pulsar PSR1257', *Nature*, vol. 355 (9 January 1992), pp. 145–7.

8. 'Habitable exoplanets catalog', Planetary Habitability Laboratory, University of Puerto Rico at Arecibo (updated regularly). Available at: http://phl.upr.edu/projects/habitable-exoplanets-catalog (accessed 14 September 2019).

9. B. Benneke et al., 'Water vapour on the habitable-zone exoplanet K2-18b', *Earth and Planetary Astrophysics* (submitted 10 September 2019).

10. *Independence Day*, written by D. Devlin, R. Emmerich, directed by R. Emmerich, Twentieth Century Fox (1996).

11. *Signs*, written and directed by M. N. Shyamalan, Buena Vista Pictures (2002).

12. *Mars Attacks!*, written by J. Gems, directed by T. Burton, Warner Bros. Pictures (1996).

13. C. Hooton, 'The *Planet Earth 2* crew put every turtle hatchling it saw or filmed back in the sea', *Independent* (12 December 2016).

14. E. M. Jones, ' "Where is everybody?" An account of Fermi's question', *Los Alamos National Laboratory CIC-14 Report Collection*, LA-10311-MS, UC-34B (Los Alamos, NM: Los Alamos National Laboratory, March 1985).

15. M. Kaku, *Visions: How Science Will Revolutionize the Twenty-First Century* (Oxford: Oxford Paperbacks, 1999).

CHAPTER NINE

1. G. Brough, 'Men who conned the world', *Today* (9 September 1991).
2. COMETA, 'UFOs and defense: What should we prepare for?', *VSD Magazine* (July 1999).
3. Alphazebra, 'Disclosure Conference, National Press Club, 27 September 2010 (extended version, English subtitles)', YouTube (8 October 2010). Available at: https://www.youtube.com/watch?v=3jUU4Z8QdHI (accessed 14 September 2019).
4. RT, '"Aliens could share more tech with us, if we warmonger less" – former Canada Defense Minister', *YouTube* (5 January 2014). Available at: https://www.youtube.com/watch?v=Pg6VTzacb9I (accessed 14 September 2019).
5. J. Carter, International UFO Bureau Inc., statement made 14 September 1973.
6. J. Miles, *Weird Georgia* (New York: Sterling, 2006).
7. 'Unidentified Aerial Phenomena in the UK Air Defence Region', Defence Intelligence Staff, internal report (2000).
8. K. Ritter, 'Area 51 events mostly peaceful; thousands in Nevada desert', Associated Press (21 September 2019).
9. B. Clinton, 'Remarks by the President upon departure', Office of the Press Secretary of the White House (7 August 1996).
10. L. Keane, 'Groundbreaking UFO video just released by Chilean Navy', *HuffPost* (5 January 2017). Available at: https://www.huffpost.com/entry/groundbreaking-ufo-video-just-released-from-chilean_b_586d37bce4b014e7c72ee56b (accessed 14 September 2019).
11. G. W. Pedlow and D. E. Welzenbach, *The Central Intelligence Agency and Overhead Reconnaissance: The U-2 and OXCART Programs, 1954–1974 (1992)* (New York: Skyhorse, 2016).
12. K. V. Whitman, 'Appeal from the United States District Court for the District of Nevada, Philip M. Pro, District Judge, Presiding', United States Court of Appeals for the Ninth Circuit No. 00-16378, D.C. No. CV-94-00795-PMP, Argued and Submitted 14 June 2002, San Francisco, California, Filed 14 April 2003.
13. G. Warchol, 'Crash site of one of Area 51's mysteries lies near Wendover', *Salt Lake Tribune* (13 June 2011).

14. A. Stolyarov, 'An experimental analysis of the Marfa lights', *The Society of Physics Students at the University of Texas at Dallas* (10 December 2005).

15. B. Dunning, 'The Brown Mountain lights', *Skeptoid Podcast*, no. 226 (5 October 2010).

16. P. Jaekl, 'What is behind the decline in UFO sightings?', *Guardian* (21 September 2018).

17. 'Humanity responds to "alien" Wow signal, 35 years later', *Space* (17 August 2012). Available at: https://www.space.com/17151-alien-wow-signal-response.html (accessed 14 September 2019).

18. D. S. McKay et al., 'Search for past life on Mars: Possible relic biogenic activity in Martian meteorite ALH84001', *Science*, vol. 274, no. 5277 (1996), pp. 924–30.

19. *The Thing*, written by B. Lancaster, directed by J. Carpenter, Universal Pictures (1982).

20. E. K. Gibson et al., 'Evidence for ancient Martian life', NASA (July 1999). Available at: https://mars.jpl.nasa.gov/mgs/sci/fifthconf99/6142.pdf (accessed 14 September 2019).

21. E. Chatzithedoridis, S. Haigh and I. Lyon, 'A conspicuous clay ovoid in Nakhla: evidence for subsurface hydrothermal alteration on Mars with implications for astrobiology', *Astrobiology*, vol. 14, no. 18 (2014), pp. 651–93.

22. E. Hunt, 'Chinese city "plans to launch artificial moon to replace streetlights" ', *Guardian* (17 October 2018).

23. T. S. Boyajian et al., 'Planet hunters X.KIC 8462852 – where's the flux?', *Monthly Notices of the Royal Astronomical Society* (17 October 2015).

24. N. Drake, 'Mystery of 'alien megastructure' star has been cracked', *National Geographic* (3 January 2018).

CHAPTER TEN

1. A. William and M. Parsons, *The Lore of Cathay: Or, The Intellect of China* (New York: F. H. Revell, 1901).

2. D. R. Williams, 'The Apollo 13 accident', NASA (12 December 2016). Available at: https://nssdc.gsfc.nasa.gov/planetary/lunar/ap13acc.html (accessed 14 September 2019).

3. 'Apollo 13 technical air-to-ground voice transcription', prepared by Test Division, Apollo Spacecraft Program Office, NASA (April 1970).

4. A. G. Stephenson et al., 'Mars Climate Orbiter Mishap Investigation Board Phase 1 Report', NASA (10 November 1999). Available at: http://sunnyday.mit.edu/accidents/MCO_report.pdf (accessed 3 November 2019).

5. *Armageddon*, written by R. R. Pool, J. J. Abrams, J. Hensleigh, directed by M. Bay, Buena Vista Pictures (1998).

6. C. Sagan, *Cosmos* (London: Abacus, 1983).

7. A. Melott et al., 'Did a gamma-ray burst initiate the late Ordovician mass extinction?', *International Journal of Astrobiology*, vol. 3, no. 55 (2004).

8. *The Day the Earth Stood Still*, written by E. H. North, directed by R. Wise, Twentieth Century Fox (1951).

9. B. Aldrin, 'We explore or we expire – it's time to focus on a great migration to Mars', *News North America* (3 May 2019).

10. E. Kim, 'Online food delivery still presents a $210 billion market opportunity', *Tech Crunch* (8 October 2016).

11. J. Kepler, *Ad Vitellionem Paralipomena* (Frankfurt: C. Marnius and Heirs of J. Aubrius, 1608).

12. 'LightSail: flight by light for CubeSats', The Planetary Society. Available from: http://www.planetary.org/explore/projects/lightsail-solar-sailing/ (accessed 14 September 2019).

13. 'Breakthrough Starshot', Breakthrough Initiatives. Available at: https://breakthroughinitiatives.org/initiative/3 (accessed 14 September 2019).

14. A. N. Shapiro, 'The physics of warp drive', Alan N. Shapiro, Technologist and Futurist, blog and text archive (14 April 2010). Available at: https://web.archive.org/web/20130424012220/http://www.alan-shapiro.com/the-physics-of-warp-drive/ (accessed 14 September 2019).

15. H. S. White, 'Warp-field mechanics 101', NTRS, NASA (2 September 2011). Available at: https://ntrs.nasa.gov/archive/nasa/casi.ntrs.nasa.gov/20110015936.pdf (accessed 3 November 2019).

16. C. Burgess and C. Dubbs, *Animals in Space: From Research Rockets to the Space Shuttle* (New York: Springer, 2010).
17. J. Chladek, *Outposts on the Frontier: A Fifty-Year History of Space Stations* (Lincoln, NE: University of Nebraska Press, 2017).
18. H. Weitering, 'NASA's Moon-by-2024 push could help put astronauts on Mars by 2033, chief says', *Space* (3 April 2019). Available at: https://www.space.com/nasa-moon-2024-landing-mars-2033.html (accessed 14 September 2019).
19. I. Sample, 'Fake mission to mars leaves astronauts spaced out', *Guardian* (7 January 2013).
20. M. Collins, *Carrying the Fire: An Astronaut's Journeys* (London: Pan, 2009).

APPENDICES

1. A. Griffin, 'Astrological signs are almost all wrong, as movement of the moon and sun throws out zodiac', *Independent* (23 March 2015).

Index